Integrated Evaluation for Sustainable River Basin Governance

Integrated Evaluation for Sustainable River Basin Governance

Edited by

Nuno Videira, Giorgos Kallis, Paula Antunes and Rui Santos

IWA Publishing

Published by IWA Publishing, Alliance House, 12 Caxton Street, London SW1H 0QS, UK

Telephone: +44 (0) 20 7654 5500; Fax: +44 (0) 20 7654 5555; Email: publications@iwap.co.uk
Web: www.iwapublishing.com

First published 2007

British Library Cataloguing in Publication Data
A CIP catalogue record for this book is available from the British Library

Library of Congress Cataloging- in-Publication Data
A catalog record for this book is available from the Library of Congress

ISBN: 1843391481

ISBN13: 9781843391487

Contents

Editors and Contributing Authors *vii*
Foreword *ix*
Acronyms *x*
Glossary *xi*

Chapter 1: Advisor Project Overview **1**
1.1. Objectives and Methodology 1

PART A: INTEGRATED EVALUATION FOR SUSTAINABLE RIVER BASIN GOVERNANCE
Chapter 2: Integrated Evaluation in the Water Framework Directive **6**
2.1. The WFD in brief 6
2.2. Evaluation in the WFD 8
2.3. Public Participation in the WFD 10
2.4. The Challenge of Integrated Evaluation 10

Chapter 3: Quality of river basin governance processes: Outcomes, patterns and requirements for future design and implementation of evaluation processes **12**
3.1. The Case Studies: Searching for Patterns 12
3.2. Patterns & Linkages 19
3.3. Contexts of Evaluation: in Pursuit of Justifications 24
3.4. Evaluation Processes: 3 pillars and 1 beam 26

PART B: PRINCIPLES AND TOOLS FOR INTEGRATED EVALUATION
Chapter 4: Framework for Integrated and Participatory Evaluations **31**
4.1. Participation in Water Resource Planning and Management: Learning from Past Experiences 31
4.2. Integrated Deliberative Decision Processes Framework 35

Chapter 5: Knowledge Assessment: Protocol for Quality Assurance by Extended Peer Review **38**
5.1. Information Flows Model 38
5.2. Quality Diagnosis: The Case Studies 39
5.3. Finding Patterns 47
5.4. Protocol for Knowledge Quality Assurance of Water Planning and Governance Processes 49

Chapter 6: Combined CBA and CEA Framework for Flood Protection **55**
6.1. Introduction 55
6.2. Description of CBA and CEA methodologies 56
6.3. Project description of the case study: Space for the River 59
6.4. Cost benefit analysis 62
6.5. Social cost effectiveness analysis 65
6.6. Introducing public participation into the proposed framework 68
6.7. Summary and conclusions 71

Chapter 7: Understanding the Value of Biodiversity in Water Ecosystems: Economics, Ethics and Social Psychology **73**
7.1. Introduction 73
7.2. Survey Design 78
7.3. Main Survey Basic Results 82
7.4. Testing the New Model of Environmental Valuation 87
7.5. Conclusions 93

Chapter 8: Scenario Workshops as a Tool for Sustainable Water Resource Governance: The Case of the Island of Naxos, Greece **95**
8.1. Introduction 95
8.2. Methodology 101
8.3. Workshop outcomes 106
8.4. Discussion 108
8.5. Conclusions 112

Chapter 9: Mediated Modelling To Support Public and Stakeholder Participation in Water Resources Planning and Management: the Baixo Guadiana Experience, Portugal 115
9.1. Introduction 115
9.2. Mediated Modelling 117
9.3. Participatory Modelling in the Baixo Guadiana 118
9.4. Results and Discussion 120
9.5. Conclusions 129

Chapter 10: Social Multi-Criteria Analysis for the Evaluation of Water Management Alternatives. The Case of the Costa del Sol (Malaga, Spain) 132
10.1. Integrated Evaluation, Multi-Criteria Evaluation and the Water Framework Directive 132
10.2. Institutional Analysis, Social Research and the Naiade Model 133
10.3. Evaluation of Alternative Options for Water Supply of the Costa Del Sol (Spain) 135
10.4. General Outcome: Social Multi-Criteria Analysis for Integrated and Participatory Evaluation of River Basin Governance 144

PART C: THE WAY AHEAD
Chapter 11: Future Challenges 150
Challenge 1 - Improving / combining deliberative tools 150
Challenge 2 - Testing IDDPs in practice 150
Challenge 3 - Improving the use of scientific information in deliberative processes 150
Challenge 4 - Increasing the decision impact of IDDPs 151
Challenge 5 - Changing governance structures 151

Annexes
Annex I: Protocol for Information used in ADVISOR 153
Annex II: Protocol for Information used in River Basin Assessment Studies 161
Annex III: Definitions of Criteria used in Quality Assurance Protocols 165
Annex IV: Problems in Contingent Valuation 170
Annex V: Population and Sample Statistics 174

EDITORS

Nuno Videira, *Ecoman Centre, DCEA-FCT, New University of Lisbon, nmvc@fct.unl.pt*
Giorgos Kallis, *Environmental Planning Laboratory, University of the Aegean, gkallis@env.aegean.gr*
Paula Antunes, *Ecoman Centre, DCEA-FCT, New University of Lisbon, mpa@fct.unl.pt*
Rui Santos, *Ecoman Centre, DCEA-FCT, New University of Lisbon, rfs@fct.unl.pt*

CONTRIBUTORS (IN ALPHABETICAL ORDER)

Paula Antunes, *Ecoman Centre, DCEA-FCT, New University of Lisbon, mpa@fct.unl.pt*
Daniel Borrego, *Ecoman Centre, DCEA-FCT, New University of Lisbon, dborrego@fct.unl.pt*
Frank Bruinsma, *Free University of Amsterdam, Fbruinsma@feweb.vu.nl*
Rob Burton, *The Macaulay Institute, University of Aberdeen, r.burton@macaulay.ac.uk*
Harry Coccossis, *University of the Thessaly, hkok@prd.uth.gr*
Serafin Corral Quintana, *University of La Laguna, scorral@ull.es*
Ângela Guimarães Pereira, *EC - JRC – IPSC, Ispra, angela.pereira@jrc.it*
Leandro del Moral, *University of Seville, lmoral@us.es*
Dionisia Hatzilacou, *Environmental Consultant, kirisis@otenet.gr*
Gary Hill, *The Macaulay Institute, University of Aberdeen, g.hill@macaulay.ac.uk*
Giorgos Kallis, *Environmental Planning Laboratory, University of the Aegean, gkallis@env.aegean.gr*
Wendy Kenyon, *The Macaulay Institute, University of Aberdeen, w.kenyon@macaulay.ac.uk*
Gonçalo Lobo, *Ecoman Centre, DCEA-FCT, New University of Lisbon, ggl@fct.unl.pt*
Alexandra Mexa, *University of the Aegean*
Pilar Paneque, *Pablo de Olavide University, ppansal@dhuma.upo.es*
Belen Pedregal Mateos, *University of Seville, bpedregal@us.es*
Rui Santos, *Ecoman Centre, DCEA-FCT, New University of Lisbon, rfs@fct.unl.pt*
Peter Shannon, *The Macaulay Institute, University of Aberdeen, p.shannon@macaulay.ac.uk*
Clive L. Spash, *Geography & Environment Department, University of Aberdeen, c.spash@macaulay.ac.uk*
Eleni Svoronou, *Environmental Consultant*
Ilias Tziritis, *University of the Aegean*
Kevin C. Urama, *The Macaulay Institute, University of Aberdeen, k.urama@macaulay.ac.uk*
Nuno Videira, *Ecoman Centre, DCEA-FCT, New University of Lisbon, nmvc@fct.unl.pt*
Ron Vreeker, *Free University of Amsterdam, rvreeker@feweb.vu.nl*
Stefanos Xenarios, *University of the Aegean*

TEXT EDITOR

Theoni Dede, *University of Thessaly, enpl@env.aegean.gr*

ADVISOR Project

Integrated Evaluation for Sustainable River Basin Governance

Advisor (Integrated Evaluation for Sustainable River Basin Governance) is a research project funded by the European Commission under the Energy, Environment and Sustainable Development RTD Programme.

Contract EVK1-CT-2000-00074

Project co-ordinator

Paula Antunes

Ecoman - Ecological Economics and Environmental Management Centre
Department of Environmental Sciences and Engineering
Faculty of Sciences and Technology
New University of Lisbon
Quinta da Torre
2829-516 Caparica
PORTUGAL

ADVISOR Partners

Ecoman Centre, DCEA-FCT/UNL, Portugal

University of the Aegean, Greece

Knowledge Assessment Methodologies
EC - Joint Research Centre - IPSC, Ispra, Italy

Free University of Amsterdam, The Netherlands

Macaulay Institute, University of Aberdeen, United Kingdom

University of Seville, Spain

Instituto da Água, Portugal

Instituto del Água de Andaluzia, Spain

A.N.E.T.K.Y., Greece

Project Documentation:

The reports and further information on the ADVISOR Project may be accessed through the Website http://ecoman.dcea.fct.unl.pt/projects/advisor

Foreword

This Report synthesizes the main results obtained throughout the ADVISOR research project ("Integrated Evaluation for Sustainable River Basin Governance") funded by the European Commission, under the "Energy, Environment and Sustainable Development" theme of the 5th Framework Research Programme.

The aim of ADVISOR was to improve the understanding of evaluation processes as part of river basin planning and management and to provide a framework supported by a toolkit for the conduct of integrated and participatory evaluations. The project comprised four work-packages.

Work Package 1 examined past water project or plan evaluation cases in five EU states and drew insights on the problems of past evaluation practices.

Work Package 2 interpreted these past experiences from different analytical angles leading to a theory for an integrated evaluation process, emphasising issues of deliberation, multiple values, quality in the use of information and governance.

Work Package 3 moved from theory to practice. Different tools which could contribute to integrated evaluation processes were tested in experimental case applications. These included scenario workshops, mediated modelling supported by a quality assurance protocol, social multi-criteria evaluation, cost-effectiveness analyses and monetary valuation.

Work Package 4 aimed at transferring the experience and lessons learned during the ADVISOR project to policy makers, contributing especially to the implementation process of the Water Framework Directive. A guidance document for designing and implementing Integrated Deliberative Decision Processes (IDDP) was developed, detailing a step-by-step procedure to achieve integrated evaluations.

The synthesis of the main results achieved throughout the ADVISOR project is presented in this report in three parts: Part A sets the stage for the role of integrated evaluations in river basin planning and management; Part B unfolds the set of relevant principles and tools developed in the project and Part C concludes with futures challenges for integrated evaluation processes.

Acronyms

ATB	*Attitude Toward the Behaviour*
CBA	*Cost Benefit Analysis*
CEA	*Cost Effectiveness Analysis*
CIS	*Common Implementation Strategy of the Water Framework Directive*
CJ	*Citizens' Juries*
CVM	*Contingent Valuation Method*
EC	*European Commission*
EIA	*Environmental Impact Assessment*
EU	*European Union*
IDDP	*Integrated Deliberative Decision Process*
MM	*Mediated Modelling*
MS	*Member States*
NAIADE	*Novel Approach to Imprecise Assessment and Decision Environments*
NHP	*National Hydrological Plan*
PBC	*Perceived Behaviour Control*
PRB	*Pilot River Basin*
QA	*Quality Assurance*
QAP	*Quality Assurance Protocol*
RBD	*River Basin District*
RBMP	*River Basin Management Plan*
SMCE	*Social Multi-criteria Evaluation*
SN	*Subjective Norm*
TPB	*Theory of Planned Behaviour*
SW	*Scenario Workshops*
VW	*Visioning Workshops*
WFD	*Water Framework Directive*
WP	*Work-Package*
WTA	*Willingness to Accept*
WTP	*Willingness to Pay*

Glossary

Citizens' Juries / Consensus Conferences	*Tools for community opinion while addressing controversial issues. They are based on the rationale that given adequate information and opportunity, all people, even if not professionally trained, can be trusted to take decisions regarded as legitimate and fair on behalf of the community.*
Common Implementation Strategy	*Multi-stakeholder approach of the EU to support the implementation of the Water Framework Directive.*
Deliberation	*A careful consideration or discussions of reasons for and against (Oxford English Dictionary).*
Deliberative decision	*A decision that involves social interaction, language-based discussion and debate, respect, reflection, and transformation of different values and viewpoints (Holmes and Scoones, 2000).*
Evaluation	*Refers to the process of assessing and comparing the relative values (in the broadest sense) of one or more courses of action.*
Governance	*Governance refers to the process whereby elements in society wield power and authority, and influence and enact policies and decisions concerning public life and sustainability. It is a broader notion than government, whose principal elements include the constitution, legislature, executive and judiciary. Governance involves interaction between these formal institutions and those of civil society (IIAS, 1999).*
Integrated Deliberative Decision Process	*A deliberative decision process consisting of one or more debates with different participants' representations, based on social robust information and knowledge; taking explicitly into account in its design and conduct the presence of multiple values and being embedded in an inclusive and discursive governance framework (institutional, regulatory, social).*
Mediated Modelling	*A tool for overcoming problems inherent in linear thinking and compartmentalized, non-participatory decision making. It aims for a collaborative team learning experience to raise the shared level of understanding in a group, as well as fostering a broad and deep consensus (van den Belt, 2004).*
Monetary valuation	*Estimation in monetary terms of the value of environmental services and goods.*
Multi-criteria evaluation	*A decision aid method based on the analysis of a number of alternatives, taking into account conflicting interests and multiple criteria, usually including economic, social and environmental factors.*
Public participation	*In broad terms refers to any of the several mechanisms (processes and tools) intentionally designed for involving lay public and/or stakeholders (the two groups together are often referred to as interested parties) in decision-making.*
Quality Assurance	*In environmental policy related issues, the maintenance of quality depends on an open dialogue among all those affected, where quality is understood as a contextual property of scientific information. The involvement of the relevant social actors (stakeholders, etc.) is suggested as a transversal element to ensure the quality of the whole evaluation procedures in river basin governance.*
River Basin	*The area of land from which all surface run-off flows through a sequence of streams, rivers and, possibly, lakes into the sea at a single river mouth, estuary or delta.*
Visioning Workshop	*Workshop which allows for creating a desired vision about the future of the examined system and generating, ranking and designing the implementation of ideas for its realisation.*

Chapter 1

ADVISOR Project Overview

1.1. OBJECTIVES AND METHODOLOGY

The process of evaluating and authorising new water related projects and plans is critical in the context of river basin management. Until now, evaluations have in many cases failed to account for sustainability concerns. The new EU Water Framework Directive (WFD) (Directive 2000/60/EC, of 23 October 2000) establishes the river basin as the geographical unit upon which resource governance may achieve its sustainability objectives, and asks for the prior evaluation and authorisation of all new river basin interventions. However, this Directive does not provide any guidance to river basin authorities on how to carry such evaluations. Unless the evaluation procedure of new plans and projects evolves into a new, multi-dimensional and multi-stakeholder participatory approach, river basin objectives as expressed in the Water Framework Directive will be at stake.

In addressing this gap of outmost importance for the sustainable governance of Europe's river basins, the main objective of the ADVISOR project has been to improve the understanding of evaluation processes, as part of river basin planning and management, and to provide a framework, supported by a toolkit, for the conduct of integrated and participatory evaluations.

In order to accomplish ADVISOR's goal, a conceptual framework for integrated evaluation was developed, consisting of three inter-related dimensions (Figure 1.1):

- *Information*: collection and presentation of the relevant scientific information to the evaluation, accounting for the quality of data, the complexity of the problem and the uncertainty of the future, and allowing for multiple description and explanatory frameworks;
- *Assessment*: applying an assessment module to the data, consisting of a comprehensive evaluation of the plausible alternative actions, accounting for all sustainability-related criteria;
- *Process*: actively involving the civil society and the stakeholders in the process of building-up the information base and performing the assessment.

None of these three dimensions may ignore the importance of the institutional, political and socio-economical context in which the river basin project takes place (*Context*).

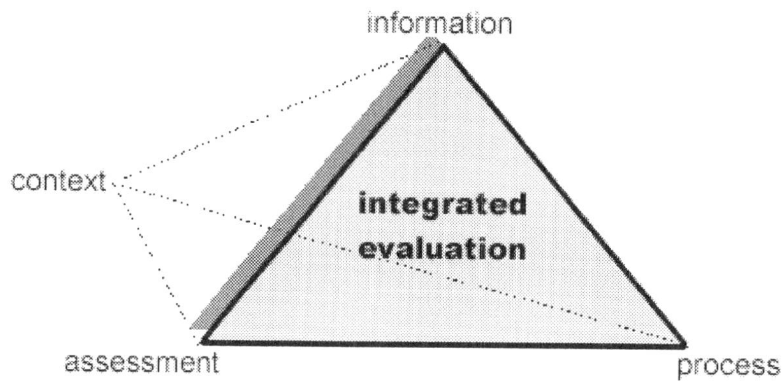

Figure 1.1 ADVISOR's conceptual framework for integrated evaluation

The development of the ADVISOR project was based on a two-tiered research structure (Figure 1.2) consisting of:

(1) The establishment of a robust theoretical understanding of the barriers and opportunities for integrated evaluation processes in a pan-European, river basin context;
(2) The delivery of a set of practical scientific and participatory methods forming an integrated evaluation framework that will help to improve the sustainability orientation of river basin project evaluation procedures.

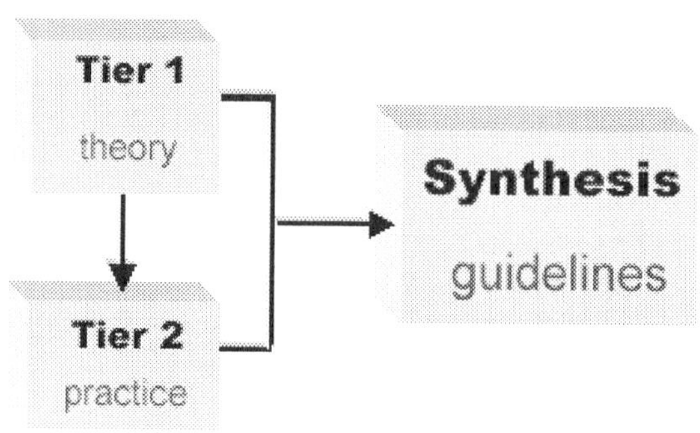

Figure 1.2 Two-tiered conceptual model of the ADVISOR project

Tier 1 was executed during the first 24 months of the project, yielding a set of reports documenting the results from the *ex-post* analysis of project/plan evaluation processes in five European countries (WP1 - Work Package 1) and the horizontal comparison of those findings (WP2 – Work Package 2) under the lens of each dimension of ADVISOR's conceptual framework for integrated evaluation.

Tier 2 comprised Work-Package 3 (WP3) and Work-Package 4 (WP4), where specific methodological tools, focussing on each of the sides of the integrated evaluation triangle of Figure 1.1, were developed and tested in selected European river basins. With WP3, ADVISOR moved from theory to practice. Different tools which could contribute to integrated evaluation processes were tested in experimental case applications. These included scenario workshops, mediated modelling supported by a quality assurance protocol, social multi-criteria evaluation, cost-effectiveness analysis and monetary valuation.

WP 4 of the project aimed at transferring the experience and lessons learned during the ADVISOR project to policy makers, contributing especially to the implementation process of the Water Framework Directive.

1.2. Key Deliverables and organization of the Final Report

In accordance with the established workplan, the research teams of the ADVISOR project have produced seven deliverables (Box 1.1), which include the several results obtained in each of the work-packages presented above.

Box 1.1 ADVISOR's key deliverables
(available from http://ecoman.dcea.fct.unl.pt/projects/advisor)

(1) Five reports on ADVISOR's *ex-post* case studies
(2) Four horizontal reports which compare the *ex-post* case studies with respect to each of the dimensions of analysis presented in Figure 1.2
(3) A meta-analysis report comparing river basin project evaluation methodologies
(4) Integrated theory presenting the outcomes, patterns and requirements for the future design and implementation of evaluation processes
(5) Five *ex-ante* case studies reports, describing methodologies and tools for integrated evaluations
(6) Integrated methodology presenting a framework for the implementation of integrated and participatory evaluation processes
(7) A guidance document on Integrated Deliberative Decision Processes for Water Resources Planning and Evaluation

This **Final Report** synthesizes the main results obtained throughout the project, organizing some of the deliverables presented above in a sequence of chapters, which combine the theory and practice developed in ADVISOR. The report is divided in three parts: **Part A** sets the stage for the role of integrated evaluations in river basin planning and management; **Part B** presents a set of relevant principles and tools developed in the project and **Part C** advances with futures challenges for integrated evaluation processes.

Chapter 2 in Part A summarizes the policy background in which the ADVISOR research has been deployed, including a brief description of the WFD and where evaluation and participation are needed within the context of implementation of the Directive. This chapter also presents the challenges for integrated evaluation within the WFD and how ADVISOR has addressed them.

Chapter 3 on "Quality of River Basin Governance Processes: Outcomes, Patterns and Requirements for future design and implementation of evaluation processes" corresponds to *Deliverable 4*. This chapter builds on the findings from Tier 1 (see Box 1.2 for a detailed list of *Deliverables 1 to 3*) to present some understanding of evaluation on river basin governance processes, and thereafter develops emergent patterns and linkages among the 4 vertices of the ADVISOR tetrahedron (Figure 1.2) and recommendations for future practice of river basin governance evaluation activities.

Part B starts with **Chapter 4** which presents the framework (including processes and tools) for integrated deliberative decision processes proposed in ADVISOR (it corresponds to *Deliverable 6*). This framework consists of a generic process on how integrated evaluations should be performed and which tools could be used at different stages of the process. **Chapter 5** on "Knowledge Assessment: Protocol for Quality Assurance by Extended Peer Review" discusses a cross-cutting issue to all tools, i.e. assessing the quality of the information used. The authors also present a protocol for quality assurance which is illustrated with an application to one of ADVISOR's case studies.

Box 1.2 Advisor's case studies and reports from Tier 1
(available from http://ecoman.dcea.fct.unl.pt/projects/advisor)

- *The Alqueva Multipurpose Project (Portugal)*, N. Videira, G. Lobo, P. Antunes, R. Santos, A. Pereira (2002)
- *The Evinos River Reservoir (Greece)*, D. Hatzilacou, G. Kallis and H. Coccossis (2002)
- *The Designation Of The River Ythan and Estuary as a Nitrate Vulnerable Zone (United Kingdom)*, G. Hill, K. Urama, C. Spash, G. Wynn (2002)
- *The Grensmaas Project (The Netherlands)*, E. van Leeuwen, J. Dalhuisen, R. Vreeker, P. Nijkamp (2002)
- *River Ebro Interbasin Water Transfer (Spain)*, L. Del Moral, B. Pedregal, M. Calvo, P. Paneque (2002)
- *Public Participation in Integrated River Basin Governance*, N. Videira, G. Lobo, P. Antunes, R. Santos (2003)
- *Quality of Information In River Basin Governance*, Â. Guimarães Pereira (2003)
- *Evaluation practices in water project decision-making processes*, G. Hill, L. del Moral, P. Paneque, B. Pedregal, C. Spash and K. Urama (2003)
- *Comparison of the Institutional Context of the five case-studies*, G. Kallis and H. Coccossis (2003)
- *Meta-analytic Comparison of River Basin Project Evaluation Methodologies,* E. van Leeuwen, R. Vreeker, J. Dalhuisen, F. Bruinsma, P. Nijkamp (2003)

Some of the tools tested in ADVISOR include: combined cost benefit analysis and cost effectiveness analysis, monetary valuation, scenario workshops, mediated modelling and social multi-criteria analysis. **Chapters 6 to 10** present a detailed description of each of these tools along with the results of its implementation in five *ex-ante* case studies (these chapters correspond to *Deliverable 5* of the project).

In accordance with ADVISOR's main findings, **Chapter 11** in Part C closes this report by presenting some of the preliminary challenges envisaged for integrated deliberative decision processes in water resource planning and evaluation, which include: improving / combining different tools, testing the full processes in practice, improving the use of scientific information, increasing its decision impact and changing governance structures.

Culminating the research developed in ADVISOR, *Deliverable 7* consists of a **guidance document** on **"Integrated Deliberative Decision Processes for Water Resources Planning and Evaluation"** (which was prepared as a stand-alone volume, also available through the project's website). It should complement the guidance developed by the specialized working groups under the WFD Common Implementation Strategy (CIS). It is primarily addressed to those authorities in the Member States (MS) of the European Union (EU) responsible for decisions in water resource management (e.g. river basin districts authorities, water utilities, and other national, regional or local agencies with responsibilities in water resource management or the authorization of water resource projects). It may also be directly useful to stakeholders and the broader public.

The platform proposed in order to achieve integrated evaluations is this of an Integrated Deliberative Decision Process (IDDP). A deliberative decision process is one in which decision is the outcome of interaction through dialogue between a broad and inclusive range of people with respect for their different viewpoints. Thus, an IDDP: combines more than one tool and participation events in a longer-term process; caters for a democratic use of scientific information; recognises the presence of multiple values; is aware of its inter-relationship with the broader socio-political ("governance") framework.

The developed guidelines aim at: creating a common understanding of IDDPs and its benefits for water resource planning and management, especially in the context of the WFD; providing guidance on how to design and implement a comprehensive IDDP for water resource planning or specific decisions; and providing tools and examples of application in order to make IDDPs operational.

Part A

Integrated Evaluation for Sustainable River Basin Governance

Chapter 2

Integrated Evaluation in the Water Framework Directive

Giorgos Kallis

Environmental Planning Laboratory, University of the Aegean, Greece

This chapter elaborates the goals of the ADVISOR project in the context of implementing the EU Water Framework Directive (WFD) (CEC, 2000). First, a brief reminder of the contents of the WFD is provided. Then, the key areas where evaluation will be needed are identified. In the third section the requirement of the WFD for public participation is emphasized. This leads to the concluding section where the new challenges to the practice of evaluation brought by the WFD are discussed. These are the challenges that the ADVISOR project comes to address.

2.1. THE WFD IN BRIEF

The Directive establishes a framework for the protection of all waters (including inland surface waters, transitional waters, coastal waters and groundwater) which (CEC, 2000):

- Prevents further deterioration of, protect and enhance the status of water resources;
- Promotes sustainable water use based on long-term protection of water resources;
- Aims at enhancing protection and improvement of the aquatic environment through specific measures for the progressive reduction of discharges, emissions and losses of priority substances and the cessation of phasing-out of discharges, emissions and losses of the priority hazardous substances;
- Ensures the progressive reduction of pollution of groundwater and prevents its further pollution; and
- Contributes to mitigating the effects of floods and droughts.

Overall, the Directive aims at achieving *"good water status"* for all waters by 2015. There are three basic classes of quality status: high (referring to undisturbed conditions or only very minor evidence of distortion), good and moderate. For surface waters the objective is that of a "good" ecological and chemical quality status. A surface water is defined as of good ecological quality if there is only slight deviation from the biological community that would be expected in conditions of minimal anthropogenic impact; a standard process is provided in the WFD for defining local standards accordingly.[1] Chemical status is classified only in two categories: "good" and "failing to achieve good". A "good" water body fulfils all the standards set by EU legislation for the concentration of chemicals.

[1] Quality elements for assessment are divided into biological, hydromorphological and supporting chemical and physico-chemical elements for rivers, lakes, transitional and "artificial / modified" waters (those created or resulting from a human physical modification and serving economic activities) (CEC 2000). The directive defines a specific process for defining harmonised, yet water body – specific parameters and standards for status.

 Integrated Evaluation for Sustainable River Basin Governance. Edited by Nuno Videira, Giorgos Kallis, Paula Antunes and Rui Santos. ISBN: 1843391481. Published by IWA Publishing, London, UK.

For groundwater the goal is a "good" status defined in terms of chemical and quantitative properties (a groundwater body can be either of a "good" or a "poor" status). A chemically "good" groundwater should comply with all existing legislative standards, should not exhibit effects of saline or other intrusions and in general, should not result in failure to achieve the goals for surface waters or in significant damage to terrestrial ecosystems (like surface waters, a process is set to define such standards upon a first list of relevant parameters). In terms of a "good" quantitative status, the criterion is that the level of groundwater is such that the available resource is not exceeded by the long-term average rate of abstraction and result in failure to achieve the objectives for surface waters, any significant diminution in the status of surface waters or damage to terrestrial ecosystems and permanent alterations to flow direction.

An important goal is this of "no-deterioration" for all waters upon implementation of the measures, or more specifically, no deterioration from a higher to a lower status. A principle of "no direct discharges" to groundwater is also established, though a number of exemptions are maintained for some common direct discharging activities (mining-related).

The goal of non-deterioration is in force since the entry into force of the WFD (22-12-2000). For the achievement of the good status, which should be achieved by 2015, a 12 years extension may be granted from the Commission after request from the Member States (MS).[2]

The "good" and non-deteriorating status is the minimum goal for all waters; in addition where more stringent requirements are needed for particular uses, "protected zones" should be established and higher objectives set within them. These should include at least areas already protected by Community legislation, i.e. drinking waters, bathing waters, nutrient (nitrate and urban w/w) sensitive designated areas and areas designated for the protection of habitats or species. In addition other zones may be designated for the protection of economically significant aquatic species and for recreational activities.

The key actions that Member States need to take include(CEC, 2000):

- To identify the individual river basins lying within their national territory and assign them to individual River Basin Districts (RBDs) and identify competent authorities by 2003 *(Article 3, Article 24);*
- To characterize river basin districts in terms of pressures, impacts and economics of water uses, including a register of protected areas lying within the river basin district, by 2004 *(Article 5, Article 6, Annex II, Annex III);*
- To carry out, jointly and together with the European Commission, the intercalibration of the ecological status classification systems by 2006 *(Article 2 (22), Annex V);*
- To make operational the monitoring networks by 2006 *(Article 8);*
- Based on sound monitoring and the analysis of the characteristics of the river basin, to identify by 2009 a programme of measures for achieving the environmental objectives of the Water Framework Directive cost-effectively *(Article 11, Annex III);*
- To produce and publish River Basin Management Plans (RBMPs) for each RBD including the designation of heavily modified water bodies, by 2009 *(Article 13, Article 4.3);*
- To implement water pricing policies that enhance the sustainability of water recourses by 2010 *(Article 9);*
- To make the measures of the programme operational by 2012 *(Article 11);*
- To implement the programmes of measures and achieve the environmental objectives by 2015 *(Article 4).*

The minimum requirement for the programme of measures is to include existing legally set instruments (e.g. secondary sewerage treatment, good agricultural practices in nitrate sensitive zones, industrial installation permits) plus a number of mandatory new instruments defined in the WFD. These include costing/pricing, zoning of designated areas, abstraction and discharge permitting and authorisation of activities that impact water quality. The WFD does not prescribe in detail these measures. The EC has set up working groups to prepare guidelines of good implementation practice (e.g. for water pricing). When the aforementioned "basic measures" do not suffice to achieve the environmental objectives of the Directive, additional measures should be implemented by river basin authorities, such as applying stricter permit standards, zones for good farming practices beyond the nitrate sensitive areas or implementing water demand management programmes.

---

[2] The wording of the text raises concerns whether MS will be held legally liable if they fail to achieve the standards. In a loose interpretation, MS are obliged only to do their best, i.e. set up the river basin authorities, produce plans and implement the measures foreseen (see below) but they can not be held responsible if the measures fail to achieve the goals.

2.2. EVALUATION IN THE WFD

Evaluation refers to the process, of assessing and comparing the relative values (in the broadest sense) of one or more courses of action. Where will evaluations be needed in the implementation of the WFD?

River Basin Plans and Programmes of Measures (Articles 11, 13, Annexes VI and VII)

Preparation of a river basin plan involves - among other - the following tasks:

- Identification of "significant water management issues" in each river basin;
- Identification of necessary additional measures in order to achieve objectives;
- Decision on a final programme of measures and appraisal of its costs.

An evaluation process is necessary for these three steps, which basically correspond to the steps of a rational planning process, i.e. problem scoping, goal setting, identification and assessment of measures and agreement in an "optimal" plan of measures.

An evaluation might also be needed concerning the benefits (including contribution to achievement of status objectives) and costs of individual measures or designs thereof. For example, after an analysis of status and issues, a river basin authority may conclude that a river restoration project is necessary. There might be various options however on the scale, technical characteristics and interventions of the project; an evaluation of the different alternatives will be necessary.

Designation of protected areas (Article 6, Annex IV)

A register of protected areas should be included in the river basin plans, including areas around drinking water sources, areas with "economically significant aquatic species", bodies of recreational waters including bathing water, nutrient-sensitive vulnerable zones and areas for the protection of habitats or species (including relevant Natura 2000 sites), where water is an important factor in their protection.

Many of these areas, their borders, standards, measures and limitations imposed are already defined as part of existing Community legislation. However, many others are still to be defined (especially areas with ecologically significant species and nitrate vulnerable zones where progress with the respective directives has been disappointing), while for many protected areas (e.g. Natura 2000 sites) plans and measures have not been finalized. Designation of these areas will need an evaluation of the advantages and disadvantages of different spatial borders, measures implemented, limitations imposed on the different activities, etc. The designation of the Ythan catchment in Scotland as a nitrate vulnerable zone (see ADVISOR's report by Hill et al. 2002) is a classic example of the type of evaluation issues faced in the designation of protected areas.

Designation of heavily modified waters (Article 4.3)

MS can designate as "artificial" or "heavily modified" waters in which the necessary hydromorphological changes for achieving good ecological status would:

(1) Have *"significant adverse effects"* on the wider environment, navigation, water storage and regulation, flood protection and other equally important *"sustainable human development activities"*.
(2) The *"beneficial objectives"* served by the artificial or modified characteristics of the water body cannot, for reasons of *"technical feasibility"* or *"disproportionate costs reasonably be achieved by other means"*, which are a *"significantly better environmental option"*.

Modified waters should generally cover regulated rivers, dams, artificial canals, etc. For heavily modified waters lower standards apply and the aim is reduced from achieving good ecological status to achieving "good ecological potential".

Judgment on the aforementioned criteria for designation will require some sort of an evaluation, for example on:

- The effects of the necessary hydromorphological changes on the environmental and economic functions served by the water body;
- Which activities can be deemed as "sustainable development" ones;
- The benefits vs. the technical feasibility, costs and environmental impacts of alternative means that could satisfy the same functions (e.g. wind farms substituting hydro-power dams).

Derogation from objectives (Article 4.5)

The Directive provides that MS may aim to achieve less stringent environmental objectives for specific bodies of water when they are so affected by human activity that the achievement of the objectives would be "*infeasible*" or "*disproportionately expensive*" and the "*environmental and socioeconomic needs*" served by such human activity can not be achieved by other means, which are a "*significantly better environmental option not entailing disproportionate costs*".

This requires an evaluation of action vs. no action costs and a multi-dimensional assessment of the polluting human activity vs. alternatives in terms of the needs served plus costs and environmental impacts.

Authorisations (Articles 4 and 11)

Controls, such as prior authorization or registration (licensing), should be applied on abstractions and impoundments of water as well as any other new activity in a river basin with "significant adverse impacts on the status of water". Although, the Directive leaves open the rules which should govern such authorizations, especially for larger waterworks (e.g. dams) there will be a need for the evaluation of their costs and benefits *vis- a-vis*. compliance with environmental objectives.

Elsewhere the Directive explicitly exempts from the requirement to respect status objectives

- "*New modifications*" to the physical characteristics of surface water or groundwater and
- "*New sustainable human activities*"

if:

- The reasons are of "*overriding public interest*" and/or the benefits to the environment and to society of achieving the objectives are "*outweighed*" by the benefits of the new modifications or alterations to human health, to the maintenance of human safety or to sustainable development, and
- The beneficial objectives served by those modifications or alterations of the water body cannot for reasons of technical feasibility or disproportionate costs be achieved by other means which are a significantly better environmental option.

In these cases it suffices to take only practical mitigation measures and to explain and justify the reasons in the river basin plans.

New dams for the drinking water supply of cities, water transfers, dams for hydroelectricity, river regulation for navigation, graveling extraction, etc, are some of the measures that could fall under the scope of this exemption. Evaluations will be needed of the scope for the modification/activity vs. environmental costs as well as a comparison in terms of costs, feasibility and environmental impact with alternative interventions. The Spanish National Hydrological Plan is an example. Note that such interventions are also subject to the requirement for an Environmental Impact Assessment and for a Strategic Environmental Assessment (in the case of large, cumulative projects).

A provisional but by no means exhaustive list of cases where evaluations will be needed is provided below:

River Basin Planning and Management	River Basin Programmes of Measures Water resource management plans (i.e. including waterworks, demand management) Individual measures (e.g. restoration projects)
Authorisations - Derogations	Designation of heavily modified water bodies Other water bodies exempted from objectives Authorisation of projects such as dams, transfers, drainage, major land-use change plans, etc
Designation of Protected Areas	Nitrate vulnerable zones Areas with economically significant aquatic species Habitats and species protected areas New recreational areas

2.3. PUBLIC PARTICIPATION IN THE WFD

Participation in the river basin planning process is a key requirement of the WFD (see Box 2.1). The Directive mandates at least the access of public to information and the duty for consultation on the part of the authorities (reporting back on how the consultation process affected the formulation of the river basin plan). The Directive however goes also one step beyond by "encouraging active involvement". Although this is not precisely defined, it is generally taken as suggesting more active processes of public involvement in actual decision-making (EC, 2002a).

Box 2.1 Requirements for Public Participation in the Water Framework Directive

Preamble 14:

The success of this Directive relies upon close cooperation and coherent action at Community, Member States and local level as well as on information, consultation and involvement of the public, including users.

Preamble 46:

To ensure the participation of the general public including users of water in the establishment and updating of river basin management plans, it is necessary to provide proper information of planned measures and to report on progress with their implementation with a view to the involvement of the general public before final decisions on the necessary measures are adopted.

Article 14 on public information and consultation:

1. Member States shall encourage the active involvement of all interested parties in the implementation of this Directive, in particular in the production, review and updating of the River Basin Management Plans. Member States shall ensure that, for each River Basin District, they publish and make available for comments to the public, including users:
   a) A timetable and work programme for the production of the plan, including a statement of the consultation measures to be taken, at least three years before the beginning of the period to which the plan refers;
   b) An interim overview of the significant water management issues identified in the river basin, at least two years before the beginning of the period to which the plan refers;
   c) Draft copies of the River Basin Management Plan, at least one year before the beginning of the period to which the Plan refers.

Upon request access shall be given to background documents and information used for the development of the draft River Basin Management Plan.

2. Member States shall allow at least six months to comment in writing on those documents in order to allow active involvement and consultation.
3. Paragraphs 1 and 2 shall apply equally to updated River Basin Management Plans.

Annex VII on River Basin Management Plans:

A River basin management plans shall cover the following elements:

9. A summary of the public information and consultation measures taken, their results and the changes to the plan made as a consequence;
10. The contact points and procedures for obtaining the background documentation and information referred to in Article 14(1), and in particular details of the control measures adopted in accordance with Article 11(3)(g) and 11(3)(i) and of the actual monitoring data gathered in accordance with Article 8 and Annex V.

2.4. THE CHALLENGE OF INTEGRATED EVALUATION

Economic valuation and cost-effectiveness analyses are obviously key tools in the process of implementing the WFD. This has been recognized by the establishment by the Commission of a Working Group on "Economic and the Environment". As the group however explains in its guidance document, economic analysis is only "a process of providing valuable information to aid decision making" (EC, 2002b). While economic valuation tools are an essential part of the overall approach for supporting decisions, they are not the only. The wording of the Directive with terms such as "other means", "significant effects", "sustainable activities", "feasibility", "disproportionate costs", "reasonably be achieved", "significantly better environmental option", "environmental and socioeconomic needs", etc, expand the range of evaluation needed and the dimensions to be taken into account. The Working Group in Economics for example recognizes that a judgment on "disproportional costs" encompasses many complex decisional, institutional and socio-economic elements … [and] ... will require co-ordination with other experts and consultation of key stakeholders and the public" (European Commission, 2002a: 35).

Multi-dimensionality in evaluation and plurality in the use of evaluative instruments are two of the challenges for an Integrated Evaluation as part of the WFD. The third challenge is the active involvement of the public. The challenge is how to move from passive consultation on predefined decisions or plans to a more active engagement of stakeholders and/or the broader public in the actual process of making decisions and evaluating alternatives. In ADVISOR this challenge is tackled by means of Integrated Deliberative Decision Processes (IDDPs) which refer to decision methods in which there is social interaction, verbal discussion and debate. IDDPs respect the different views of participants and aim to the transformation of the values and preferences held. While the goal is to reach good and effective decisions, deliberative processes perceive negotiation and participation as containing value over and above the quality of the decisions that emerge (Holmes and Scoones, 2000). IDDPs benefit from the use of the state-of-the-art decision support aids (e.g. models, softwares, G.I.S, multi-criteria methods, etc) and combine them with processes of facilitation (workshops, conferences, etc). Hence, they provide a good "platform" upon which to integrate evaluation with active engagement of the participants.

The other challenge of evaluation in the WFD is how to deal with complexity and uncertainty in information. The necessary scientific information upon which to base WFD decisions is often missing (lack of monitoring systems, etc) or is unreliable. Many hydrological or pollution effects are extremely complex and have a stochastic or probabilistic nature. Furthermore, different actors may have different sources or analyses of information.

The ADVISOR project comes to address these challenges at both a theoretical and a practical level. It provides an integrated theory of evaluation suggesting criteria to appraise the quality of an evaluation process in the context of the WFD and it develops tools both for the economic valuation and for deliberative evaluations, which can be used by river basin authorities in their implementation tasks.

REFERENCES

Council of the European Communities (CEC) (2000) Directive of the European Parliament and of the Council: Establishing a framework for Community action in the field of water policy. 2000/60/EC, (23 October 2000), *Official Journal of the European Communities*, L 327/1.

European Commission (2002a) *Guidance on public participation in relation to the water framework directive. Active involvement, Consultation and Public access to information,* Directorate for Environment: Brussels.

European Commission (2002b) *Economics and the Environment. The implementation Challenge of the Water Framework Directive. A guidance Document,* Directorate for Environment: Brussels.

Hill, G., Urama, K., Spash, C., Wynn, G. (2002) *The Designation of The River Ythan and Estuary as a Nitrate Vulnerable Zone*, Report prepared for the WorkPackage 2 of the ADVISOR project, United Kingdom.

Holmes, T. and Scoones, I. (2000) Participatory Environmental Policy Process: Experiences from North and South, Institute for Development Studies (IDS), Working Paper 113.

Chapter 3

Quality of River Basin Governance Processes: Outcomes, Patterns and Requirements for Future Design and Implementation of Evaluation Processes

Ângela Guimarães Pereira

EC - JRC – IPSC. Ispra, Italy

This chapter will develop some understanding of evaluation on river basin governance processes, based on past experiences and thereafter develop emergent patterns and linkages among the 4 vertices of the ADVISOR tetrahedron (context, information, assessment and participation) and requirements that could be recommended for future practice of river basin governance evaluation activities. It looks for justifications of past practice on 4 types of context: institutional, societal, knowledge and methodological. Eventually, it suggests a framework for quality of evaluation of river basin planning and governance processes, which is further developed in Chapter 5.

3.1. THE CASE STUDIES: SEARCHING FOR PATTERNS

In this section a summary of the comparative analysis of the 5 case studies carried out according to the ADVISOR tetrahedron is done: **Context** based on the document done by Kallis and Coccossis, 2003; **Information** based on the document done by Guimarães Pereira, 2003; **Assessment** based on the document done by Hill *et al.* 2003; and **Participation** based on the document done by Videira *et al.* 2003.

3.1.1. Context

Analysis of context (Kallis and Coccossis, 2003), took into consideration **institutions[3] involved and main social actors** in the evaluation processes. The authors performed a horizontal analysis of institutions and the other vertices of the ADVISOR tetrahedron, for the 5 case studies. The remaining of this section summarises this analysis.

[3] A thorough review of meanings of "institutions" is available in Kallis & Coccossis (2003).

 Integrated Evaluation for Sustainable River Basin Governance. Edited by Nuno Videira, Giorgos Kallis, Paula Antunes and Rui Santos. ISBN: 1843391481. Published by IWA Publishing, London, UK.

3.1.1.1. Institutions & assessment

"There seems to be a distinction between the Ythan and Southern countries case studies and the Dutch case. For the former cases, water institutions, formal and informal, are directed to the satisfaction of productive concerns. The rational can be described as sectoral goals getting translated into political demands and waterworks being built according to the sectoral demand. Hence, basically the evaluation of the latter are based on political criteria, plans and planning being merely instrumental for works programming rather than actual strategic choices of alternatives that could possibly respond to several goals.

In the Dutch water institutional system, there seems to be high level of integration through "function" and "area-based" approaches, joint plan preparation and implementation through project organisations for specific water interventions and the duty for integrated (water, natural, spatial-physical) planning at the level of the province.

> *In terms of the process and timing, the Dutch case is closer to the ideal of an integrated evaluation than the other four cases and this is in accordance with the relatively more progress towards an integration of the institutional structure of the water sector in the Netherlands".*

The EC regional policy emphasises that integrated evaluation should be viewed as a means to integrate developments in the national and regional water policies and decisions. That is, for instance, at the basis of the Strategic Environmental Assessment initiative.

"In the Ythan case, the EC environmental regulation cannot be simply taken as a case for enforcement of quality standards but rather a more fundamental effort for institutional (and property rights') change, shifting the distribution of costs and benefits. More specifically, the nitrates' directive, based upon the 'polluter pays' principle, shifts the cost of the cleaning up agricultural pollution from the public and the State to agricultural producers. As the authors of the Scottish case note (Hill et al. 2002),

> *this marks a fundamental change with respect to the prior regime of property rights, whereby producers were 'free' to use their land as they wish and be paid by the State if they were to implement environmental protection measures. The new institutional structure, as shaped by the nitrates' directive, and the fact that it is coming from a supreme (in terms of institutional and sanction powers), higher level authority (the EC) allows the Scottish government to pursue this change, despite the reaction it will reasonably meet from farmers who will see their income being threatened."*

"The case of Greensmas also presents a case of implementation of the 'polluter pays' principle and a shift in property rights. Income from gravel extraction is to be used for the development of a new natural area and for a flood protection scheme."

"In the Evinos case, there are valid claims that the project was not the response to the drought crisis, but that the very crisis was 'socially constructed', in order to pave the way for the project and the much-needed 'absorption' of EC support funds. This is also hinted in the Alqueva and the Ebro cases, where EC funding retains central role in the realisation of the projects. This is not implying that the projects were conceived because of EC funding. All three were long-held and conceived 'strategic' public infra-structure projects. The high economic cost though (relative to the potential of the country economies) in an era of austere public budgets, may partly explain the delay in the implementation of these projects. Availability of EC regional funding must have been central in the decision to go forward with these long-held, 'strategic' projects of national importance."

"EC funding may be a mechanism that taps opportunities for actual strategic reflection."

3.1.1.2. Institutions & participation

The following are the issues identified in the 5 case studies with regards to the role of institutions in participation of wider sectors of the society in the planning or decision processes of the water projects reviewed here:

- *Institutions did not promote public strategy debates in 4 case studies (Southern and Ythan); the institutional relation is rather of informative or consultation natures. There is no regulation explicitly requiring that when project is in early stages, as well as no [spontaneous] initiatives from relevant institutions to do so. Participation is not embodied in the institutional structures that govern decisions at strategic level;*
- *Except for the Dutch and UK cases, participatory processes, often carried out during the decision phase; timing jeopardised effective contribution to the implementation of the project;*
- *When participation occurred, discussions were often partial, being not on the real decisions, which is left to the policy spheres;*
- *No rules set by the institutions as to who and how to participate (e.g. as in the Ythan case). Institutional arrangements governing participatory exercises do not seem to be addressed for meaningful involvement of relevant parties;*

- *No surprise then that as echoed in all Advisor case-study reports, public participation have remained to a large extent superficial and used as mechanisms of justification of an already taken decision rather than real instruments of public engagement in the spirit of an integrated evaluation;*
- *It does not seem sufficient to prescribe or mandate participation or participatory institutionalised mechanisms. Institutions governing participation and access to decision process are usually framed by those who are already in control of decisions, and therefore there is an inherent bias in the preservation of the status-quo. The Spanish government, for example, would be careful not to change the composition of the Water Councils and institutionalise majority representation from 'new actors', as this would be equivalent to welcoming the rejection of the NHP. Similarly, when governments, the main "developers" of public works themselves, are asked to "determine the public concerned ... and ... specify the way in which the public may be informed ... and consulted" (CEC, 1985), they can in theory structure a 'participatory process' that could silence opposition to their plans;*
- *The Dutch case study seems to convey the idea of being more in line with inclusionary governance styles.*

3.1.2. Information

Analysis of quality of information was based on flows of knowledge (data, tools, communication to the public, etc.) during the evaluation process: framing, assessment and public participation. These were reviewed on grounds of quality categories adapted from Corral Quintana (2000). An extended review can be read in Chapter 5 in this book, whereas the full review is available in Guimarães Pereira (2003).

Information used in the 5 case studies is of all sorts: quantitative time series, model estimations, probabilistic or fuzzy information, etc.), qualitative data (nominal, pictorial, etc.), in several formats and from several sources. The remaining of this section summarises main findings in relation to quality of information used, activities undertaken and style of communication for public goals.

3.1.2.1. Quality of Data Used

→ *Fitness for purpose*

- *Adequacy and Relevance: in general for all studies the information used seems to be specific to the case study or generic to the issues covered*
- *Accuracy: all studies report high scientific uncertainties and inconclusive nature of scientific evidence*
- *Completeness: all case studies seem to be bound by great omissions of information. There are two situations:*
    - *Omission of data is recognised within the assessment studies: the strategy is usually to call for further studies (all case studies use this strategy) or use data/results from other similar studies*
    - *Omission is NOT recognised within the assessment studies but viewed by as such by peers and other relevant social actors (e.g. Alqueva case study)*

→ *Applicability*

- *Access & Availability: access to available information for assessment purposes does not seem to be a problem in any of the studies.*
- *Intelligibility: information for assessment purposes seems to be highly specialised, essentially intelligible for peers, in the majority of the cases*

→ *Reliability – CONTROL*

- *Sources of information (WHERE): Scientific Literature; Research project reports; Institutional publications (governmental and others); Expert/scientific consultancy – very little if any information is coming from social research.*
- *Sources of information (Methods used): Calculations; Estimations (scenarios, model simulations, etc.); Historical values/time series/etc.; Field work/measurements.*
- *Sources of information (Generators): Mainly experts and consultants*
- *Verification & Colleague consensus about sources: In many cases information used and produced is controversial (e.g. Ebro case study)*

→ *Reliability – CONFIDENCE*

- *Extended peer acceptance: Information in many cases has a low degree of acceptance by the non-scientific communities*
- *Legitimacy: Information coming from peer reviewed scientific literature seems to be legitimate; reports suggest sometimes that technical studies done for the assessment are not credible".*

3.1.2.2. Quality of Activities Undertaken

This section refers to procedures, methodologies, models, etc. used in the assessment process.

→ *Fitness for purpose*

- *Adequacy: some reports suggest that some of the methods or conceptual approaches were NOT adequate: some of the expressions used are, e.g. conceptual errors, overestimations, omissions, underestimations, wrong assumptions, inadequacies...*
- *Adaptability: some studies have deployed models which may be generic – but it is not clear if they are indeed (e.g. adoption of US EPA models in European cases); in some cases, methodologies seem to have been tailored or purpose built for the case studies (e.g. Alqueva case study).*
- *Transformation/encoding: large range of transformation types – extrapolations; aggregations; etc. Rarely data is used as collected (raw). There is no indication by reviewers that has been used in a manipulative way as such, just misinterpreted (e.g. Alqueva case study).*

→ *Transparency*

- *Documentation: usually sources are stated, but less usually though are explanations, justifications or statement of assumptions made in relation to models, methods, etc used in the assessment communicated through documents aimed at non-specialised audiences*
- *Arbitrariness: see above.*

→ *Legitimacy*

- *Colleague consensus: in many cases the methodologies used were not consensual among peers".(e.g. Ebro case study).*

3.1.2.3. Style of communication

→ *Intelligibility*

- *Compliance with audience: reports suggest that the studies done are suitable for technical and scientific audiences; in many cases, it is not clear whether the assessment reports are suitable for non-scientific audiences, including policy makers. Some assessment studies (e.g. developed within EIAP) require the availability of a non-technical reports or consultation documents and so, in principle these were the sources compliant with non-expert audiences. Also, many times the press and propaganda material were the sole intelligible sources of information for non-experts, yet it is not surely the desirable or effective means for informing wider involvement of relevant communities.*
- *Accessibility: Some assessment studies (e.g. developed within Environmental Impact Assessment Processes (EIAP)) require the availability in specific locations. Internet was also used to make reports available (e.g. Alqueva and Ythan case studies). In terms of accessibility it must be noted however, that in many cases the time available for consultation, participation, etc. was not satisfactory (all cases).*

→ *Transparency*

- *Documentation: difficult to appreciate this issue with information provided in the reports; it seems as though that sources of information were clearly stated, but there is little basis to conclude that models' assumptions, justification, applicability conditions or necessary adaptations, were available at all.*

→ *Communication of uncertainties*

- *Statement of uncertainties: In some studies acknowledgement of uncertainties (including ignorance) and major gaps of information is done. For some studies uncertainty is not an issue.*
- *Implications for the decisions stated: This seems to be lacking for the majority of the cases. In the majority of the cases, decisions were made, despite uncertainties. The Precautionary principle is largely ignored.*
- *Pedigree of results: there are no such types of information available; however, in some cases other studies or sources are referred to.*

In general dealing with uncertainties throughout the studies is treated in the following way:

- *Not an issue*
- *Suggestion for further studies*
- *Go ahead despite uncertainties: implications for the formulation of the problem and for the assessment phases are not clear*
- *Precautionary principle seems to be largely ignored (in the Ythan case study it is suggested that in-action is adopted as a precautionary measure)*

Therefore, since the quality of the information used in the studies will reflect on the quality of the decisions and outcomes, it emerges from this evaluation that a different type of control has to be implemented. Indeed in order, to have relevant information from the societal point of view quality assured extended quality assurance have to be implemented which can be achieved through extended peer review processes with those concerned".

3.1.2.4. Information from actors into the process

"Rarely there seems to be any type of consultation for the formulation of the problem or the definition of scope of the assessment studies (exception is for instance the Alqueva IEIS, where an external panel was invited to determine the scope of the study). It seems as though that post-reviews are more favoured than ant-reviews.

When the participatory process occurs, it is not clear how recommendations are integrated in the assessment studies and/or decision processes – and actually whether they are at all considered in the process (evaluation and decision). In some cases, however, concerns from public hearings have been incorporated into mitigation measures and when controversy was high, s actual conflict took place, the whole appraisal process being jeopardised (example: Ebro and Grenmaas case studies). Yet, this is not genuine public participation process, but a public mobilisation one.

In most cases there seems to be just a one-way information flow: information from the proponents of the projects towards the public. In few case studies there are suggestions that actual consultations took place, the results of which are far from being normative and recommendable (e.g. no formal ways of incorporation of the consultation outcomes, timings prohibitive for serious involvement, etc.)".

3.1.3. Assessment

This section is entirely based on the analysis done in (Hill et al. 2003). It examines the different stages of assessment for the case studies with respect to five specific elements: **tools, criteria and values, timing, responsibility and influence**.

"1. Four of the case studies (Alqueva reservoir-Portugal, Evinos reservoir-Greece, Ythan NVZ-UK and Ebro River Transfer-Spain) have been formally and definitively approved by the responsible administrations (national governments in all cases) and are in a more or less advanced implementation phase. In most of these cases, this fact does not exclude the existence of some important uncertainties as regards to the real feasibility of the projects or likelihood of their complete implementation.

The fifth case considered (The Grensmaas-The Netherlands) has reached a stalemate situation due to disputes concerning some important aspects of the project (quantity of gravel that should be dredged), lack of support by the inhabitants and legal allegations. The intended planning timing has changed and the project is so far in a "preparation phase" that have had to be fulfilled between 1997 and 1999. Furthermore, the direction of the project, i.e. organisation Maaswerken, had to resign from their tasks in January 2002. This situation could be considered similar to some stagnation stages experienced by other projects, e.g. Alqueva reservoir between 1978 and 1980 or Ebro transfer between 1995 and 2000. Nevertheless, the stalemate situation currently affecting the Grenmaas project, together with other distinctive features which will be commented further below, could be eloquent of some substantial differences in the governance issues encompassed by The Netherlands case.

2. Three projects (Alqueva, Evinos and Ebro) are characterised by a strategy of 'productivist' management of water resources, not exempt of strong tensions and substantial changes throughout the formulation and decision process. In any case, the projects have as central objective, and as main driver behind the whole decision making process, the generation of water resources to supply urban, industrial and/or agricultural demands. This, together with other factors from the geographical context, gives them remarkable similarities. As the report on Evinos points out: "It seems that the supply management approach, seeking to satisfy ever growing demands, prevailed against a demand management approach, the main weighing factors being the pressure of time, the urgency of the matter, the anticipated EU subsidies and perhaps the knowledge and safety inherent in the implementation of tested technologies (dam constructions), instead of new innovative ones."

The Scottish case study (Ythan) can be regarded as a representative case of a 'post-productivist' approach to water environment management, as far as it aims to protect or recover the quality of this environment, reducing the intensity of economic activities (agriculture) that are supposed to have a negative incidence over them. This fact introduces substantial differences in the nature of the contents that feature this project's decision making process, i.e. legitimacy and hegemony (Bourdieu 1977, 190-197[4]) of values and interests at stake, position of actors' networks, coherence with the general trend to 'naturalisation' dominant in the surrounding cultural context, even though the four projects already approved (Alqueva, Evinos, Ebro y Ythan) have other highly significant common characteristics.

The fifth case study (The Grensmaas) begins as a 'multifunctional project'. However, one of its initial objectives (gravel recovery, although in a less aggressive manner than traditionally carried out) is increasingly contested by social demands of protection and recovery of water environment and landscape quality, which constitutes the other basic initial objective. Furthermore, the increasing weight of environment and landscape conservation contradicts the objective of financial balance, through gravel exploitation, explicitly included in the formulation of the project. In this

[4] Bourdieu, P. (1977): Outlines of a Theory of Practice, Cambridge.

same case, over the decision making process (mainly since 1994-1995 floods) an already ongoing revision of the general strategy to deal with flooding has got strength. The new schemas (from controlling river flows to living with them, i.e. "Space for the River") become superposed and should be compatible with the project's initial objectives.

3. Regardless of the specific objectives behind them, the four projects formally approved (Alqueva, Evinos, Ythan and Ebro) coincide in a key feature that is firmly established in the formulation of the project and appears clearly stated in the respective case study reports: the assumption that the benefits are greater than the costs.

In some cases (Alqueva, Evinos, Ebro) this assumption is based on widely accepted and hegemonic social values and in long held traditions of hydraulic policy in the history of each country. In Portugal, a strategic water reserve and the regional development of Alentejo ("The hypothetical social benefits (e.g. regional development) of the Project were the main arguments to proceed with it despite the fact that economic benefits (irrigation, energy and water supply) aren't clearly demonstrated and the environmental impacts of the Project are very significant due to the submersion of a very large area". In Greece, overcoming a water deficit that prevents the development of the main national metropolis by implementing a project foreseen decades ago, an unquestionable objective reinforced by the situation created by a specific drought spell. In the Spanish case the solution to hydrological imbalances which constitutes the country's main geographical problem. This does not mean that these strategies are immune to internal tensions within the water policy arena and to changes in the wider cultural, political and economic contexts that frame the decisions adopted at national level. Actually, these assumptions are deeply rooted in the kind of socially hegemonic 'schemas', 'shared strategies' and 'myths' about water environment, well described for other contexts by Social Theory (Anderson 1980, Thompson, Ellis et al. 1990, Burke 1992, 101-103[5]).

In a similar way, the legal framework (EU Nitrate Directive) of the Scottish case (Ythan) implicitly implies that benefits of reducing nitrate pollution in an area designated as an Nitrate Vulnerable Zone (NVZ) will justify any costs that may be incurred in achieving these reductions. As opposed to the before-mentioned cases, given the 'postproductivist' nature of the project's objective, social legitimisation of this assumption demands changes in the configuration of power balances and of power relations between the different networks of stakeholders, i.e. a transition from the hegemony of the alliance between Ministry of Agriculture and Fisheries (MAFF) and the National Farmers Union (NFU), in the context of a nation state, to the emergence of a more pluralistic process of policy development where wider interests are no longer marginal and subservient to the MAFF-NFU alliance, with Scottish Wildlife Countryside Link (SWCL) establishing connections in Brussels, in a sub-state/ supra-state context. Changes in actors and in decision-making scales, inevitably interconnected among them, which express and accompany transformations in social objectives and values.

On the other hand, the Grensmaas project is based in the assumption that "no changes will occur in the markets for sand and gravel". This leads to a lack of interest on prospects about the impact of extra gravel and sand on prices of these resources, thus conditioning the cost-benefit analysis. Nevertheless, it seems that the Grensmaas case does not present a clearly hegemonic, in a social and/or legal way, assumption that could be compared with those present in the other cases, in terms of basic and strong driving-force of the project.

4. When they are present, the assumptions about the unquestionable value of the project are based on an ***indisputable diagnosis*** *of the problem and its solution, and on the indisputable definition of the necessity and technical viability of the required infrastructures or programs: 1. Regional underdevelopment open to change through hydraulic intervention, in the Alqueva dam project; 2. Stagnation due to water deficit of the country's main development area open to solution by implementing a historical project, in the case of Evinos River reservoir; 3. Surplus and deficits between regions that are necessary and possible to be balanced by means of inter-basin transfers, in the Spanish National Water Plan; 4. Eutrophication processes that can be solved through changes in agricultural methods, in the Ythan River.*

In the Grensmaas project despite the importance of this above-mentioned assumptions about market stability, the initial goals of the project (gravel exploitation together with protection against floods by means of canalisation and dikes) were already weakening when the project was being launched, and generate even deeper doubts during the decision making process. Thus, diagnosis and definition of alternatives have been more controversial.

5. All these factors lead to the following characteristics of the assessment carried out over the decision making process. They are more or less clearly present in the different cases according with their specific features as were presented before.

a) Underlying the overall process are the deeper issues of ***values and beliefs*** *influencing both perceptions of nature of the problems to be tackled, as well as of optimal ways to manage them. In all the cases, the high importance of established patterns and beliefs can be confirmed, in the way that has already been stated in other contexts: "Assessment of risks and response options tends to follow, rather than lead, political target setting, and the range*

[5] Reference available in Hill *et al.* (2003): Burke, P. (1992): *History and Social Theory*, Policy Press, Cambridge.

of options tends to contract over time" (Committee on Global Change Research 1999, 318[6]). Formal institutional procedures are not neutral but embody beliefs and ideas (informal institutions) that provide an advantage to some actors over others, acting as independent or intervening variables between the preferences and power of actors on the one hand, and condition closely policy outcomes on the other (Jordan 2000).

b) ***Assessment*** *(be it strictly economic or multi-criteria oriented) is not really considered as an independent, less say previous, stage in the design of the projects. Evaluation, beyond the partial yet firm diagnosis of the problem and its solutions, is done after the formulation of the project or even after its approval and it is carried out, if at all, incompletely.*

c) The ***implicit strategy,*** *strongly assumed in the formulation of the project, makes difficult if not impossible the discussion of global alternatives. In the two case studies can be confirmed that "policy is often strongly path dependent in that early decisions may constrain or determine later ones, thus making discussion of alternative policies extremely difficult at later stages" (Gooch et al. 2002[7]).*

d) The ***complexity*** *of ecological and social processes inherent to the projects tends to be simplified by the starting diagnosis, basically incomplete. As has been concluded in other cases (Nilsson and Langaas 2002), more effort appears to be put on collection of information on state and impacts than on driving forces, pressures and responses. The diagnosis is conceived as a sum of non-contextualised representations and supposed to be objective and existing independently of human agents, following what Thsoukas recognises as an information reductionism (Thouskas, 1997: 832[8]).*

e) The ***uncertainty,*** *present in basic aspects of information about the relations among different components of natural and social systems, is masked by the certainty that accompanies the diagnosis of the problem and the definition of the strategic solutions adopted. Decision-making processes based on the grounds of scientific alone criteria fail to recognise the complexity and the high levels of local scientific uncertainty and lead to a highly subjective and 'political' outcome.*

f) Evaluation, conditioned by this fundamental fact, constitutes a ***basically informal and iterative process****, sometimes diffuse, throughout the entire project life, from conception to completion, even over the post-decisional stage. In this changing process, the modules or vertices concerning actors' participation, available and really used information together with the context that conditions the whole decision making process, are difficult to separate from evaluation. This does not exclude the formalisation of specific evaluation documents, in different forms and rhythms, sometimes late over the project formulation or even once it has been approved. This last fact is quite indicative of the real role that sometimes evaluation performs throughout the process, i.e. an instrument for the justification and defence of an already decided strategy.*

g) Nevertheless, in general some ***evolution and development of assessment methodologies*** *and techniques can be found, closely depending on the time extension of the decision making process and the intensity of the social controversy. This, in certain cases, leads to significant changes influencing the dimension or complementary aspects of the project, although not affecting its core contents.*

h) Evaluation is closely dependent on the balance of power among actors (Mann 1986, 1, 518-521[9]), habitually affected in the case studies by an ***increasing polarisation****. To conclude from this that the decision making process in the cases outlined above were just 'subjective', 'arbitrary' and/or 'groundless' may be considered a simplification. The evaluation item plays, precisely, a key role in the way conflictive approaches are expressed and at times partially solved. In reality, "achieving effective water governance based on sound decision making process (including water projects assessment), is inherently political in nature" (Hall, 2002). We can say that the integrated perspective and practice that comprehends complexity of water issues' evaluation is inseparable from explicit considerations of ethics and policy (Funtowicz et al. 1996[10]).*

i) In the context of intertwined and partially overlapping tiers of authority that characterise the emerging system of multi-level governance, the ***European Union scale****, for legal and/or financial reasons, performs a key role as a driving-force for the emergence of formalised evaluation inputs throughout the decision making process."*

---

[6] Committee on Global Change Research, B.O.S.D., Policy Division (1999): Global Environmental Change, Washington, D.C., National Research Council.

[7] Gooch, G. et al. (2002): *Policy formulation, management and implementation in the transition to social, economic and environmental sustainability*, Mantra-East project, Mid-term Meeting, Tartu (Estonia).

[8] Tsoukas, H. (1997): "The tyranny of light. The temptations and the paradoxes of the information society." *Futures* Vol. 29. nº9, pp. 827-843.

[9] Mann, M. (1986): *The Source of Social Power*, Cambridge.

[10] Funtowicz, S., M. O'Connor and J. Ravetz (1996): "Emergent Complexity and Ecological Economics" in Van del Straaten and Van den Berh (Ed. 1996), Economy and Ecosystems in Change, Island Press.

3.1.4. Participation

This section is based on report from Videira *et al.* (2003) the analysis considering the following aspects: **compliance with legal and institutional dispositions**; **timings** in the process; ***representativeness*** (relevance of those involved); **information provided**; type of **involvement**; **outcome** or impact in the project or subsequent action.

Process design

(1) General objectives of participation: compliance with regulation or seeking for compromise solutions;
(2) *"participation impact: with exception of the Dutch case, participation has failed to promote high impacts on decision process"*;
(3) Types of parties: mainly governments on one hand and NGOs on the other (usually on different sides of the terrain, with exception for the Scottish case). In the Netherlands, the community seems to be *de facto* represented. On the whole, for most of the cases seemingly stakeholders with vested interests were not part of the process. EC played mostly a role on financing and enforcing.
(4) Timing: public involvement in the review phase of the evaluation processes, took place only in regularly stipulated consultation moments; mostly there was no contribution at a strategic level. Indeed, involvement often only made in compliance with regulation and in some cases, was a sham from the beginning, e.g. Greek case. Only the Dutch case seems to have followed a desirable structure: early involvement where relevant issues could be identified.

Implementation

The forms of implementation have been mainly public hearings and public comments; there seems to be no organised or efficient ways of getting people involved to usefully contribute with exception of the Dutch case.

Involvement took place mostly in late stages of the process. As there are no purposeful organisation of the participatory process, the necessary information deployed or gathered from such processes is not dealt with in organised forms either.

Assessment

(1) Outcomes: only the Dutch case produced some impact on the decision process; others contributed only to the mitigation plans or did not have any impact at all.
(2) Achievement of social goals: the Dutch process indicated that the decision process can be improved when early involvement takes place, where contested terrains can be dealt with and some sort of shared ground can be achieved. Exclusion of interests would intensify conflicts.

Final comments

"Legislation exists but practice does seem to correspond; there is free interpretation and poor monitoring of responsible institutions; desirably, the very process should monitored by relevant sectors of the civil society. Indeed, practice set mostly to comply with regulation being mostly irrelevant since contribution is nll."

Finally, it must be noted that the 3 projects where participation seems to be irrelevant and not timely have their origins on 30's, 50's or 60's, before regulation called for public involvement of any sort; whereas the other two cases initiated during the 90's (Grensmaas and Ythan).

3.2. PATTERNS & LINKAGES

This section seeks to explore linkages on patterns from the evaluation exercises of the five case studies, among the 4 vertices considered in ADVISOR: context, information, assessment and participation, exploring the linkages among them, in particular the influence of contextual conditions on the other 3 vertices and the information and knowledge management on assessment and participation. Finally, the role of participation in assessment is also explored.

3.2.1. Context dependency issues

Context seems to reflect in the patterns found for all vertices. Context understood as ***socio-political framing***, ***institutional arrangements*** and ***regulatory provisions***.

We start by describing patterns that show context influence on the organisation of river basin evaluation processes and how it shapes practice and outcomes.

- With exception of the Netherlands, ***political systems are highly centralised***, which shows on ***hierarchical, top-down and formalised management practice***, including for water management. This is based on the hydraulic and engineering paradigms, many times "artificially" integrated in regional development plans, but often responding to specific interests.

- Except for the NL and UK case studies, where regional authorities are quite powerful, the main actors appearing in all case studies have ***strong vested interests*** belonging to sectors such as ***engineering, agriculture, construction companies, utilities (electricity, etc.); hydraulic bodies***. Their lobbying activities are quite evident, many times fostering ***socially constructed crisis*** in order to serve particular lobbies. In fact, except for the Spanish and the Dutch cases, the contexts in which the projects were proposed were not, in first place, surrounded by public controversy. It must be noted that for those latter cases a "bottom up" approach was adopted, in which consultation with identified stakeholders was adopted. Controversy arose in a later stage; on the Spanish case a national movement against the PHN, lead by academics and NGO's; and on the Dutch case the project eventually was not approved, after an extended process of evaluation.
- The dynamics for project proposal and evaluation seems to be also quite similar in all cases: from *sectoral* goals with water demands, through water works and eventually evaluation (political, economical, technical, environmental/social). Regulatory setting, in particular that imposed by ***EC, pulls a set of evaluation demands, often acting as guardian of different values through regulation, financial support and external arbitration***, having in many cases a perverse effect on the development and implementation of the projects. ***Funding institutions seem to have a potential role to enforce evaluation***.
- As many of the ***evaluation processes*** (especially those focused on environmental and social aspects) are ***legitimacy rubber stamps*** for project approval, it is not surprising that evaluation culture is in most cases, based on ***mitigation and/or adaptation measures***.

The remaining of this section identifies context dependency in terms of the other three vertices considered in ADVISOR: information, participation and assessment.

Information

Information is the basis of evaluation, deployed in assessment and participation procedures. Two different flows can be identified, information that is deployed in the several design and evaluation phases and information that is produced as outcome of those procedures. The context in which evaluation takes place has determined in all cases the quality of information used and the outcomes of assessment and participatory procedures. Also, the very same context determines the quality of information, i.e. what information and knowledge fit and/or are relevant for each evaluation phase, completeness, accuracy, etc. requirements. The following are aspects common to the ADVISOR case studies:

- There seems to be ***no provision in the national legislations*** regarding water governance to reflect upon ***quality of information***, in any phase of the evaluation processes; at best this is an issue that is left to those who perform the assessment, while deciding on data sets and sources or analytical methods. The ***precautionary principle or extended consultation*** can, in some cases be interpreted as a political attempt to deal with ***uncertain*** information.
- Since there is not in the majority of the cases, involvement of an extended community at a strategic level of the evaluation process, ***no quality check*** (i.e. regarding compliance to wider concerns of relevant societal sectors) is performed on issue framing, scope of the assessment, information or methods deployed, etc. Moreover, in none of the cases is there ***any institutional requirement for advice on framing or scoping to relevant wider sectors of the society***, what is discouraged by hierarchical, centralised socio-political structures.
- Decisions are taken in ***contexts of uncertainty, arguable assumptions or poor justifications***, for at least 4 of the case studies.
- Five <u>principles</u> underpin good governance as proposed in the European governance White Paper: *openness, participation, accountability, effectiveness and coherence*. "*Each principle is important for establishing more democratic governance*." → the 5 case studies analysed earlier, in one way or the other do not correspond for the most part to these pillars of good governance.
    - Openness is about institutions working in an open manner and in terms of information that is, providing understandable information to all concerned: "[institutions] *should use language that is accessible and understandable for the general public*". → From the analysis done earlier, there is a need for procedures that tailor relevant information deployed in the assessments, to become intelligible and accessible to all concerned, in order to overcome the divide between institutions and other sectors of the society.
    - Participation is about inclusionary processes: "The quality, relevance and effectiveness of EU policies depend on ensuring wide participation throughout the policy chain – from conception to implementation." → The analysis done earlier confirms that active involvement of those concerned is the way to create shared ground on issues that concern all; it is the tuning mechanism of issues framing, assessment activities, etc. by the relevant community.

- Accountability is about institutions being transparent for their duties and take responsibility for what they do. Accountable processes are clear in terms of responsibility and therefore are likely to generate more trust. → The analysis done earlier confirms that accountable river basin governance processes potentially enhance trust on institutions involved and create better conditions for concerted action among all actors of a basin. This encompasses transparency in information production and processing, and every time seeking for consensual, legitimate procedures.
- Effectiveness refers to timely and at the appropriate level decisions and action. → The analysis done earlier confirms that in river basin governance, and in particular as far as information is concerned, this means that if consultation and therefore review or scoping activities will be carried out, they should be done timely to be effectively taken into account,
- Coherence refers to "*policies and action* [which] *must be coherent and easily understood*". → Incongruence is the synonymous of mistrust. The analysis done earlier confirms that coherence must start with framing of issues, where adopted reasoning is explicit and assumptions clearly justified. That is also dependent on the quality of information used and especially on the adequate balance of perspectives, different types of knowledge and value framing.

Participation

Participatory activities respond to regulation specific requests. Since, on one hand, each case study underwent a different evaluation process and, on the other hand it was developed in quite different time scales, going across different political *momentums* (national and European), participation is also framed in different ways, and contributes differently to the evaluation process itself. It must be noted that only from the middle 80's, public participation conceived as "involvement" at same stage of the project appraisal, has been considered in a regulatory act of the EU (De Marchi et al. 2001). However the following patterns are observed:

- participation ***cannot be considered in any of the cases a genuine exercise of co-production of knowledge; there is not such a concept as seeking involvement to improve the quality of the evaluation process***;
- participatory activities are ***held to comply with regulatory frameworks*** that govern the evaluation process; very often the ***articulation of the participatory activities with the whole evaluation process is inadequate and inefficient*** in view of a desirable interaction from those involved and those who perform assessments;
- in some cases (the Dutch and Spanish cases) participation is ***sought to try to avoid conflict or to address existing conflict***; in fact, in some of the cases, participation was instrumental in sought of legitimacy or conformity;
- except for the Dutch and somehow the Scottish cases, participation is seen mostly as a ***one way information delivery***; hence, the types of ***participatory methods used are inadequate to stimulate serious engagement of the interested parties*** in the process and develop agency; ***guidance*** is missing, the most common method being "public hearings" which are very often uncomfortable forums for participants and place the "providers of information" as the experts and holders of all relevant knowledge and the others as the naïf claimers that cannot articulate their (legitimate) claims in ways that the promoters of the process could understand and process…
- although regulation exists, the institutional arrangements and formal processes for involvement are still immature, in many cases ***resources being the nadir of the participatory activities*** and timing a typical excuse; this also shows that ***participation has little political interest***;
- regulation lacks provisions for ***guidance on formal or established methods*** to identify participants or "relevant stakeholders" – in some cases representativeness seems to be poor, being an object of conflict and controversy; some cases suggested that relevant social actors are somehow "forgotten" or it was not clear whose decision was to involve participants;
- Finally, ***institutional arrangements seem to be stiff to accommodate in daily practice extended evaluation processes***, the reason being perhaps the hierarchical structure, centralisation of water management issues. It would be easy to say that institutional arrangements do not hold a structure to accommodate extended processes, but that is far too simple an assertion: given that some public involvement is already taking place, it would not be too difficult to adapt the endeavour to be in greater correspondence with inclusionary governance styles. This suggests that another of the key issues for the irrelevancy of participation in evaluation processes is the fact that ***institutions are not actually prepared for what to do with the outcomes, being unable to profit from the communities' resources***.

Assessment

In the following paragraphs the influence of contextual conditions in assessment activities for the 5 case studies is outlined.

Assessment activities, in particular in what regards environmental and social aspects, are carried out to comply with regulation of project appraisal and the main pattern is that they were done after project approval or in advanced stages of it.

- ***"Environmental impact assessment studies" or flavours of it are the most common*** requirement in the case studies (exception for the UK), but Cost Benefit Analysis types of studies are also commonly used. Although EIA studies have been the compulsory assessment frameworks in Europe since 1985, for projects that could pose a threat to the Environment, their implementation is far from being prone to controversy, given their scope. ***From the studies done in the ADVISOR cases the most "integrated" assessment tools of all are EIA***.
- The history of the project plays a role in the assessment procedures adopted, as well as in the outcomes. It is interesting to note that in 3 case studies (Alqueva in Portugal, Evinos in Greece and the Ebro in Spain), the projects' first "stones" are all before the 70's (1956, 1964 and 1933). Hence, ***it is not surprising that framing has subsequently changed or if not changed, not shared, throughout the different political and economical momentums***. Naturally, the number of evaluation studies available for these three projects is also greater than for the other two more recent projects, hopefully reflecting those changes. Indeed, it would be expected that project scoping would have acknowledged these changes, yet what is actually seen is that ***there were problems of comprehensiveness***, despite the number of studies.
- Sometimes assessment does not seem to help with stated goals, when there are hidden agendas with regards to the actual objectives of the project. For instance, the Alqueva project was first conceived to deal with demands from the agricultural sector, assessments being done with that objective in mind; however, the Alqueva "mission" has been reformulated with time, its implementation being strongly decided upon political strategy that had little to do with the regional development as it has been framed in the last decade; assessment however was carried out having as ***referring objectives*** the latter ***politically correct objectives***. When assessment is not done in correspondence to a stated framing, the result is that some incongruence is found as to the assessment and decisions made and what is worse, it leaves a sense of missed opportunity and questioning about the project's usefulness and relevance.
- A common pattern in the studies is that evaluation seems taking place mostly missing a broad picture, even if some studies claim to be strategic or plans. For instance, the irrigation plans for *Alentejo* region based on the Alqueva dam seem to be outdated if one takes into consideration the EC agricultural policy.

3.2.2. Information in Assessment & Participation

Information quality and management is another key aspect that influences evaluation activities.

Assessment

To describe the patterns found, the structure proposed in Hill et al. 2003, is borrowed partially, information being understood here as data, tools and methods deployed in the assessment phases of the project. Assessment is done on the basis of the problem framing formulated mostly as a result of economical or political objectives, including compliance with regulation; framing determines in principle information and knowledge requirements and quality and the quality of information (accuracy, social robustness, legitimacy, etc.) used to frame the issues will then determine the quality of the whole process of evaluation.

- In four of the cases, EIA type of studies were applied in the assessment phase, although they are inherently multi-disciplinary, ***the outcomes are far from reflecting "integrated perspectives"*** desirable in new governance schemes and much explored in academic spheres. Integrated assessments are characterised by dealing with systems complexity, plurality of values and uncertainties, etc. And whilst this is strongly dependent on the quality of accessible information, it is also strongly dependent on the ***problem framing and*** on the ***scoping of the assessment***. These in all cases were ***essentially set by political opportunity or lobbying***.
- ***Arbitrariness seems to be a common thread of the tools deployment in the case studies***. It has different forms, like poor justification of its usage, poor fitness for purpose and tools serve often as legitimacy devices of decisions taken ahead. Also, there are issues of transparency and legitimacy, for instance in the Spanish case study, where there could not be a peers review since the authors of the technical studies were not known.
- ***Uncertainty is either not dealt with, masked by usage of crisp methods or in a cautious perspective claiming for further studies or data collection***, which is the closest to a sort of uncertainty management far from being accountable.

- In three of the cases the ***most important outcome of the assessment activities are the mitigation measures***; in the Alqueva project, for instance the volume of mitigation measures is almost as large as the assessment part itself. That suggests that a negative impact statement would be ignored anyway, and that ***there is not an a priori intention to stop the project if the assessment recommends so***.

Participation

This section addresses information provided for the participatory activities and information gathered through these activities.

As participation seemed in all cases to be a mere act of compliance with existing regulation, it is in a sense "controlled" by established lobbies at decision level; basically it is not expected to give anything, but a sense of "harmony" and consensus formation.

- The EIA processes require a "non technical document" to be accessible to all the public, the full EIA being also available for consultation in specific places. These ***documents were generally available and accessible, still in some of the cases jargon loaded, with no care for communication aspects***.
- It can be said that access to the ***information provided to the public reflected often the quality of the studies***, and because the relevance of the assessment was often not compliant with the concerns of societal sectors, it is not strange that ***information available was limited, often irrelevant for the participatory activities***;
- this probably also explains why involvement (understood here as ***spaces for exchange of information and knowledge***) is poor and when it exists it is in the form of one-way flow of information to unprepared participants to actually engage in a debate, irrelevant for the ***agendas of those who promote the participatory act which do not accommodate a serious, non-discriminatory exchange of information and knowledge***.
- What was missed in most of the cases was the belief that participatory outcomes can be useful for the whole process in terms of information brought, other than just mechanisms of justification or pseudo conflict resolution. ***With few exceptions (namely the Dutch case study) the participatory mechanisms did not inform effectively the evaluation activities***.

3.2.3. Participation and assessment

As stated earlier, there is no such thing as thinking of involvement as enhancing the quality of the assessment and even less of co-production of knowledge.

- Occasionally, involvement is used to resolve conflict and so, to discuss ***compensation or mitigation measures***.
- Except for the latter Alqueva EIA study, which involved for the scoping steps a panel of experts, the rule is that ***involvement is not used for scoping at all and even less for framing***.
- And, except for the Dutch case, ***participatory activities occurred in later stages of the evaluation process***, there was no public strategic involvement, discussions were partial, almost irrelevant in terms of ***information brought to the assessment and hence inputs cannot even prove to be useful***, more likely if they were done in early stages of the project appraisal.
- Involvement is not generally effective, not articulated with the several phases of the process; hence its usefulness is not expected. ***Timing is either an appropriate excuse for non-involvement or an actual organisational problem***.
- Finally, the use of participation as a step to inform the assessment stages, suffers from installed procedural problems and from institutional status quo.

3.2.4. Summary

The diagnosis obtained from the analyses done around the 4 main ADVISOR vertices: context, participation, information and assessment activities, on past river basin governance processes allowed the identification of types of unwanted outcomes, as well as success stories, insights onto justifications to adopt certain procedures and linkages between information and context and the other vertices of the tetrahedron. Some patterns could be found, namely the bad quality of knowledge used in many cases: observations were mostly dealing with a great deal of uncertainties, irrelevant information and most of all incomplete knowledge about relevant issues. In addition, there was little room to include knowledge other than technical and scientific in the assessment procedures and linked to the poor quality of information used there was a standing problem with poor, incomplete and arbitrary assessments, limited scopes and eventually unattended effects are reported. In addition, conflict seems to be a common pattern arising from exclusionary governance processes or late involvement of those concerned in the processes. Finally, very often, there was a poor correspondence of the governance processes with normative and legislation drivers of new governance styles (directives, white papers, etc…).

3.3. CONTEXTS OF EVALUATION: IN PURSUIT OF JUSTIFICATIONS

In this section we try to reflect on the analysis and diagnosis provided in the other two sections, aiming at providing insight onto the possible justifications that lay past evaluation of river basin projects. We set this reflection on what we identified as contexts of evaluation: institutional, knowledge, societal and methodological. These four contexts set in a sense the types of justification that determine the quality of the evaluation processes. Although their analysis will be done in separated sections they overlap and are intertwined.

Institutional context

Historical legacy

The historical gap in which the (chronologically speaking) older projects developed (Evinos reservoir, Alqueva dam and Ebro *interbasin* transfer) plays an essential role on the characteristics of the evaluation process. Some of the case studies developed under autocratic regimes, where dialogue was restricted, there were heavy hierarchical institutional structures and water management and project assessments were bound by technocratic values and approaches. Moreover, historical analysis provides insight into the changes of framing, respondent to changing national and international economical and political strategies. In a sense because these projects were long standing political promises, and there was no political courage to abandon them or make major reformulations in view of different societal or political contexts, what is reformulated is the types of justifications to have the project going ahead.

Hence, wordings like integrated perspectives, *inclusionary* governance, extended framing and scoping, acknowledgment of uncertainties, quality of information, plurality, etc. were not part of the inspiration context of planners and decision makers at the time those projects were conceived.

Institutional arrangements

Three of the evaluation processes of river basin management instances in ADVISOR have operated under governance styles nowadays arguable. First, as we have seen centralised structures make it difficult to actually frame the governance issues at an adequate scale, the proposals for action often being far from the integrated development of the river basin and its communities, but rather corresponding to political objectives or nationwide ones (that was the case of Alqueva and Ebro).

Lack of specific competence to deal with public involvement jeopardises the usefulness of the process, including timely organisation of the event, purposeful creation of spaces of exchange of knowledge and incorporation of outcomes in the assessment phases.

Institutions in Europe are in a sort of transition, to accommodate new perspectives on governance. After a number of trust crisis, there is a sort of process where tentatively those in charge of planning try to collect inputs in the field, mostly for political reasons, seeking legitimacy, and seeking conflict resolution. In one way or the other there seems to be a tendency to have more inclusive forms of governance than in the past, as the Dutch and UK cases demonstrate.

Regulation and guidance

Nowadays there is a great deal of legislation in Europe that regulates and guides environmental evaluation (EIA: CEC, 1985; CEC, 1997 and CEC, 2001c; WFD: CEC, 2000; public participation in environmental assessment: CEC, 2003) and also specifically for water issues. In relation to public participation, for instance, provisions do exist in principle, but they are in many cases quite ambiguous as to its implementation leaving space for *smart* arbitrary implementations in terms of timing and methods of involvement, as well as information exchanged and reflection on the evaluation process. Hence, lack of operational guidance may be wasting the institutional endeavour to accommodate participatory procedures in their practice, and make the results useful for the various stages of the evaluation process.

Water function

Until quite recently Water was considered only as a highly politicised production factor, from which stems the hydraulic and engineering paradigms in order to manage it! Nowadays, its social, environmental and cultural functions are recognised (for instance, in the Dutch case, water function was considered) and that inevitably leads to changes in the evaluation processes and its governance.

Societal context

In recent years governance styles have started featuring more dialogue, in an attempt of calling wider sectors of the society to be accountable in decisions about complex situations, where relevant knowledge may be limited, inadequate or uncertain. In these cases we have seen that when involvement is effective it is because it is essentially linked to vested interests, lobbies or attempts to resolve conflict.

It is obvious that those who participate determine outcomes; institutions will tend to maintain *status quo* so that decisions taken are not argued (Kallis & Coccossis, 2003). Added to this, as we have seen before, institutional arrangements are naturally resistant to change, rigid and more willing to conform to installed practice. It is believed that timing of the participatory steps could change this; except for emergency planning situations, river basin governance and planning should not be constrained by tight timings; when they exist they are probably related to political momentum, including government programmes or funding opportunities.

The absence of a purposeful institutional infrastructure that is specifically in charge of organising societal involvement seems to be one of the reasons for the lack of articulation and poor results of participation exercises. Participation is an organised event, deploying specific methodologies, very much like any other evaluation task, and should be practiced as any other step of the evaluation process. It is not just enough to have regulation allowing the "publics" to participate and cannot be let to societal initiative and self-organisation. That is a naïf assumption and, when self organisation occurred (rather called public mobilisation) in the case studies analysed, it arose from serious conflict and anger from interested parties, leading to disruption of dialogue and *concertation*.

Apart from poor correspondence with normative and legislation drivers of new governance styles (directives, white papers, etc…), the styles of governance in which the evaluation processes operated were exclusionary, which created situation of intense conflict and crisis of societal trust, in some cases culminating on disruption.

Methodological context

The types of tools deployed in the studies are bound on one hand by the types of vested interests associated with the projects and on the other hand by the regulatory framework. The studies associated with river basin management are typically technical and economical (CBA) studies and EIA studies, the latter being the most "integrated" of all. In general the assessments performed were poor, incomplete, arbitrary, based on inadequate scopes and resulting on unattended effects.

Uncertainty management reflects a typical approach in political and scientific spheres until recently, which consisted of using either by reducing or by ignoring it, instead of acknowledging and stating it, acting on precaution. The legitimacy in which science and science practitioners have been operating in the past is also hold by professional consultancy that develops the assessment activities. The technocratic and *reductionist* methodologies and methods deployed are those shared among professionals and policy circles. They tend to ignore the societal context in which they are applied and no wonder they tend to emphasise economical aspects in pursuit of sound justifications to rubber stamp policy options. Probably, some guidance about tools and methods could improve the assessment contexts.

We believe that most methodological problems of the evaluation processes can be easily resolved, methodologically speaking – see methodlogical Chapters in this book. In view of more integrated assessments, there are a number of formal and "informal" methods that can deal with complex, multiple perspectives and values, through which alternative projects can be assessed and compared. In other *problematiques*, such as climate change, where systemic relations are complex, facts are uncertain and stakes high, integrated assessments have proved useful at least to have a systemic view and promote an informed debate about options for action. That is in our belief the most important benefit from assessment tools, to allow knowledgeable debates and choices.

Knowledge Context

Knowledge deployed in many of the cases was unfit for the purpose, prone to different types of uncertainty, controversial, or unavailable. Apart from expected gaps of scientific and technical knowledge, that might have to do with unavailability of historical data, inconclusive research, inadequate formats (aggregation, scale, etc.), other types of problems arise, for instance, property rights, poor articulation (including lack of regulation) of those who perform assessments and the providers of information. Added to this, quality of knowledge suffers also from the reliability and legitimacy of sources and peer acceptance of processing tools and methods.

On the other hand, if there are no mechanisms that ensure the articulation of knowledge coming from different sources into the assessment phases (framing, assessment activities, communication) the knowledge context cannot be expected to be plural, holding multiple perspectives and being socially robust as vented in inclusive governance.

3.4. EVALUATION PROCESSES: 3 PILLARS AND 1 BEAM

River basin governance and planning is essentially a contested terrain, characterised by complex situations (many conflicts worldwide are due to water governance issues), values in dispute, high stakes and surrounded by different types of uncertainty. These are features of what Funtowicz & Ravetz (1990) described as post-normal situations, which require a different type of framing that they called post-normal science.

Hence, we would argue that in order to achieve good quality river basin evaluations at the light of new styles of governance, a number of principles be respected. As mentioned earlier, we have formulated those as 3 pillars and 1 beam (Figure 3.1), as the basis to attain a socially robust (Gibbons, 1999) framework for evaluation of river basin governance:

- Extended peer review
- Robust Knowledge
- Inclusive governance
- Transparent assessment procedures

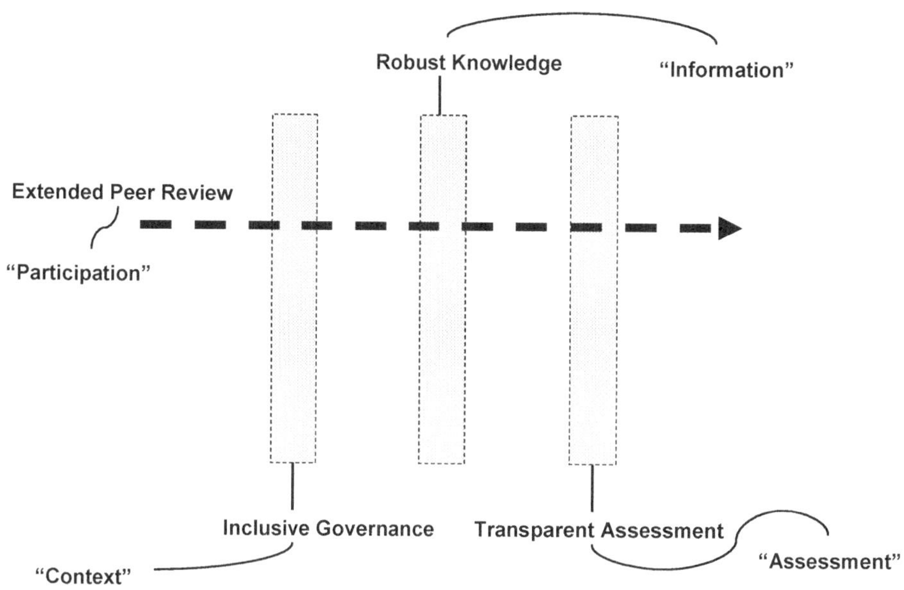

Figure 3.1 3 pillars and 1 beam, the basis for quality river basin evaluations

Extended Peer Review

In environmental policy related issues, the maintenance of *quality* depends on an open dialogue among all those affected, where *quality* is understood as a contextual property of scientific information (Funtowicz and Ravetz, 1990; Ravetz, 1996; Funtowicz, 2001). By *extended peer review* it is meant the involvement of an *"extended peer community"*, consisting not merely of persons with some sort of institutional accreditation, but rather of all those with a desire and commitment to participate in the resolution of the relevant issues. Hence, the involvement of the relevant social actors (stakeholders, etc.) is suggested here as a transversal element of the whole quality evaluation procedures in river basin governance. This is intrinsically related to all pillars of this framework and that is why we set extended peer review as a beam that crosses all pillars.

Extending decision processes, entails the creation of conditions to identify, to involve and to engage the relevant community – imply methodological and institutional aspects; the creation of mechanisms of reflection of participatory' outcomes into the evaluation process – imply methodological and institutional aspects; and to encourage society to access and to further these attempts of change of democratic rights – imply societal and institutional aspects. This is the realm of participatory activities.

Participation is an organised process; first, the spaces whereby social actors interact with process have to be created and accommodated in institutional practice; institutional bodies should know what format is best to have useful, meaningful, effective dialogues, appropriation of the issues and competence development, since that requires the knowledge of the context in which the exercise takes place. However, it is not up to the institutions to decide who participates BUT it is up to the institutions to organise the context in which the participation takes place and ensure that relevant parties can participate. Secondly, institutions have to create *ears*, i.e. mechanisms by which relevant outcomes are incorporated in the evaluation process. The realisation that the involvement is relevant and effectively contributes to produce robust knowledge about the evaluation process is a condition of trust, essential to further societal involvement.

Timing and period of involvement: little time for involvement means little time to cope with gaps, late justifications, pretence extended review and questionable incorporation of consultation outcomes into the review process and final decisions – this seems to be quite acknowledged by the provisions of WFD's Article 14 (CEC, 2000).

Inclusive Governance

Recent years have seen a growing institutional interest in governance *issues* that can affect or be affected by a relevant extended community. Partly stemming from an agenda generated by the 1992 United Nations Conference on Environment and Development, accountable, inclusive processes have been progressively encouraged over the last decade. In its White Paper on European governance (CEC, 2001b), the European Commission emphasises the need for a revised relationship between existing institutional arrangements and civil society – see Box 3.1. Recent cases related to health and safety are part of the justification for the urgency of reform, but discussion on the need for change in institutional arrangements and relationships has been ongoing for some time. Specifically, commentators have identified a need for constructing new processes of decision-making, including new structures and new modes of thinking, communication, and interaction between parties. As a result, there has been a very significant increase in opportunities for citizens' involvement in decision-making. European Union legislation is beginning to reflect this conceptual re-definition of governance relationships by modifying regulations to shift the form of entitlement from a right to be informed, to a right to participate (De Marchi *et al.* 2001). This is what it is referred here as more inclusive governance, one of the pillars of good quality evaluation of river basin governance.

Box 3.1 Five principles underpin good governance as proposed in the European governance White Paper (CEC, 2001b): ***openness, participation, accountability, effectiveness and coherence. "Each principle is important for establishing more democratic governance."***

- Openness is about institutions working in an open manner and in terms of information that is, providing understandable information to all concerned: "[institutions] *should use language that is accessible and understandable for the general public*
- Participation is about *inclusionary* processes: "*The quality, relevance and effectiveness of EU policies depend on ensuring wide participation throughout the policy chain – from conception to implementation.*"
- Accountability is about institutions being transparent for their duties and take responsibility for what they do. Accountable processes are clear in terms of responsibility and therefore are likely to generate more trust.
- Effectiveness refers to timely and at the appropriate level decisions and action.
- Coherence refers to "*policies and action* [which] *must be coherent and easily understood.*

Robust Knowledge

Robust knowledge, in the sense of knowledge that fits the purpose should not be circumscribed to scientific quality but to the actual fitness to context and relevancy to the social actors concerned. Robust Knowledge is another pillar for the quality of river basin evaluation.

In order to produce socially robust knowledge (Gibbons, 1999), *extended peer review*, i.e. involvement of all those concerned in the quality assessment process (Funtowicz, 2001) is needed in several phases:

- Extended scoping: it became obvious from the studies that governance of river basin is dependent on the type of information produced during assessment activities and its relevance, adequacy, intelligibility and legitimacy for the actors of concern. Extended scoping aims at genuine coverage of issues, so that assessment is not just a means of meeting prior commitments.
- Extended review: this corresponds to current consultation processes. Their effectiveness entails issues such as information accessibility, intelligibility and legitimacy, timely mechanisms of consultation, formal mechanisms of processing consultation outcomes into the assessment processes, yet another aspect of creating interfaces between society and policy.

A knowledge assessment protocol is desirable as part of the formal requirements of the evaluation process used in normative (guidance) and descriptive fashions. That is a protocol to assess the quality of information used and methods that process it should be applied to all phases of the process, namely framing, assessment and public involvement activities. The protocol should specifically address the pedigree of the information used, along the same methodological lines deployed in this report.

Also, specific provisions to create interfaces for mediation of the knowledge produced to the potential users and or processes. Information used for consultation has to correspond to other dimensions of quality, namely it has to be suitable for the potential audiences. Therefore, non-technical summaries as they are currently produced are not enough. There is a need for specific establishment of procedures for delivering scientific and technical knowledge to non-expert audiences.

Last but no least, guaranteeing quality of information and methods deployed in river basin assessment activities is the condition *sine qua non* for effective dialogue among all actors of a basin. Otherwise, the danger is to shift the debate from the governance issues to quality of the assessments done.

Transparent Assessment

Transparency is one of the pillars of new styles of governance (CEC, 2001b), and this should be visible throughout all processes related to policy implementation and decision-making. Transparent assessments entail explicit assessment procedures (black box free), supported by good quality information. Transparency is an essential condition to create trust, and as with information quality, assessment outcomes are to be the focus of debates about river governance, and not be questionable about their legitimacy.

The desirable tools and methods to improve the quality of assessment activities are those that can capture the characteristics of river basin governance, namely their complexity, multiple values and perspectives, different vested interests, uncertainties of different types, etc. These tools have to be able to address different knowledge representations, different languages or articulation of arguments, systemic and trans-disciplinary integration and essentially they should allow the necessary dialogues on options for river basin planning and governance. Many of these tools were promisingly tested in the ADVISOR work – see the remaining Chapters of this book.

FINAL COMMENT

The 3 beams and the pillar constitute a good starting basis to ensure the quality of river basin evaluation processes. Alone, they are just yet another good idea; to be effective and respond also to the good ideas of the WFD (CEC, 2000), they have to be *operationalised* in a series of practical recommendations.

ACKNOWLEDGEMENT

I would like to thank Lola, António and Ana for taking care of little Nicolau while I was wrapping up this chapter.

REFERENCES

Council of the European Communities (CEC) (1985) Council Directive of the 27th June 1985 on the assessment of the effects of certain public and private projects on the environment, *Official Journal of the European Communities*, C **175**, 40-49.

Council of the European Communities (CEC) (1997) Council Directive of the 3rd March 1997 amending Directive 85/337/EEC on the assessment of the effects of certain public and private projects on the environment, *Official Journal of the European Communities*, L **73**, 5-15.

Council of the European Communities (CEC) (2000) Directive 2000/60/EC of the European Parliament and of the Council establishing a framework for the Community action in the field of water policy, Published at *Official Journal (OJ L 327)* on 22 December 2000.

Council of the European Communities (CEC) (2001a) Democratising expertise and establishing scientific reference systems, Document of 2/07/2001, Available at: http://europa.eu.int/comm/governance/areas/index_en.htm.

Council of the European Communities (CEC) (2001b) European Governance: A White Paper; Com (2001) 428, Brussels, 25.7.2001, available at: http://europa.eu.int/comm/governance/white_paper/index_en.htm.

Council of the European Communities (CEC) (2001c) Council Directive of the European Parliament and of the Council on the assessment of the effects of certain plans and programmes on the environment, *Official Journal of the European Communities*, L **197**, 30.

Council of the European Communities (CEC) (2003) Council Directive of the European Parliament and of the Council providing for public participation in respect of the drawing up of certain plain and programmes relating to the environment, *Official Journal of the European Communities*, L **156**, 17-24.

Corral Quintana, S. (2000), *"Una Metodología Integrada de Exploración de los Procesos de Elaboración de Politicas Públicas"*, Universidade de La Laguna, Tenerife, ISBN 84-688-4314-8.

De Marchi B. Funtowicz, S. & Guimarães Pereira, Â. (2001) From the Right to Be Informed to the Right to Participate: Responding to the Evolution of the European Legislation with ICT, *International Journal of Environment and Pollution*, **15**(1), 1-21.

Del Moral, L., Pedregal, B., Calvo, M. and Paneque, P. (2002) River Ebro Interbasin Water Transfer, Work Package 1 of project ADVISOR, Universidad de Sevilla.

Funtowicz, S. & Ravetz, J. (1990) *Uncertainty and Quality in Science for Policy*. Dordrecht: Kluwer Academic Press.

Funtowicz, S. (2001), Peer review and quality control. In *International Encyclopedia of the Social & Behavioral Sciences,* ed N. J. Smelser & P. B. Baltes, pp. 11179-11183, Oxford: Pergamon.

Guimarães Pereira, Â. (2003) Quality of Information In River Basin Governance Review of Case Studies: Information Vertex, Work Package 2 of project ADVISOR, EC: Joint Research Centre.

Gibbons, M. (1999), Science's new social contract with society, *Nature* 402: C81-C84.

Hatzilacou, D., Kallis, Giorgos and Coccossis, H. (2002) The River Evinos Reservoir, Work Package 1 of project ADVISOR. University of the Aegean.

Hill, G., Urama, K., Spash, C. and Wynn, G. (2002) The Designation of the River Ythan and Estuary as a Nitrate Vulnerable Zone, Work Package 1 of project ADVISOR, The Macaulay Institute.

Hill, G., Del Moral, L., Paneque, P., Pedregal, B. Spash, C. and Urama, K. (2003) Evaluation Practices in Water Project Decision making Processes: cpomparative analysis of Alqueva Dam (Portugal), Evinos Reservoir (Greece), Ythan NVZ (UK), The Grensmaas (The Netherlands) and Ebro River Transfer (Spain), Work Package 2 of project ADVISOR, University of Seville and Macaulay Institute and University of Aberdeen.

Kallis, G. & Coccossis, H. (2003), Comparison of the Institutional Context of the 5 Case Studies, Work Package 2 of project ADVISOR, University of the Aegean.

Van Leeuwen, E., Dalhuisen, J., Vreeker, R. and Nijkamp, P. (2002) The Grensmaas Project, Work Package 1 of project ADVISOR, Vrije Universiteit Amsterdam.

Videira, N., Lobo, G. Antunes, P., Santos, R. and Guimarães Pereira, Â. (2002) Alqueva Multipurpose Project, Work Package 1 of project ADVISOR, New University of Lisbon.

Videira, N., Lobo, G.. Antunes, P. and Santos, R. (2003) Public Participation in Integrated River Basin Governance, Work Package 2 of project ADVISOR, New University of Lisbon.

Part B

Principles and Tools for Integrated Evaluation

Chapter 4

Framework for Integrated and Participatory Evaluations

Nuno Videira[1], Giorgos Kallis[2], Paula Antunes[1] and Rui Santos[1]

[1] *Ecoman Centre, DCEA-FCT, New University of Lisbon, Portugal*

[2] *Environmental Planning Laboratory, University of the Aegean, Greece*

Part B of the ADVISOR Final Report includes the results from the project's Work Package 3, which aimed at the development and testing of methodologies and tools for integrated and participatory evaluations within the context of river basin planning and management. This chapter introduces PART B by presenting a framework (integrated methodology) to implement such evaluations which is this of an Integrated Deliberative Decision Process (IDDP). The first section summarizes the lessons learned in the first cycle of ADVISOR's research with respect to the participatory methods and tools deployed in past evaluation processes and discusses the rationale leading to the proposal of the IDDP framework. The final section presents a generic process for integrated evaluations, with a summary description of the steps included in the proposed IDDP framework.

4.1. PARTICIPATION IN WATER RESOURCE PLANNING AND MANAGEMENT: LEARNING FROM PAST EXPERIENCES

Water resource planning is a rational, staged process of selecting and implementing the best combination of water management measures and instruments in order to achieve a set of defined goals. A generic planning process includes several tasks which start with a scoping of the problems to be addressed and collection of data to appraise conditions. This is followed by the definition of the goals that the plan should fulfil the elaboration of the goals into a set of criteria upon which to assess alternative options and the identification of alternative management options/measures. Alternatives are then comparatively evaluated upon the identified criteria leading to the compilation of an optimal mix of measures and instruments in a programme or action plan. Continuous collection of data and monitoring of the results of the measures accompany the plan's implementation. New problems and needs arise, leading to a new planning cycle (Kallis et al. 2004).

Evaluation and comparison of the advantages and disadvantages of the different management options is a crucial stage of the planning process. In reality many water management decisions are taken outside a formal planning process. Authorities have to decide whether to authorize or not a new water abstraction or dam, proceed or not with the

 Integrated Evaluation for Sustainable River Basin Governance. Edited by Nuno Videira, Giorgos Kallis, Paula Antunes and Rui Santos. ISBN: 1843391481. Published by IWA Publishing, London, UK.

restoration of a polluted river, designate or not an area as protected, or how to design new water tariffs. Ideally, all these decisions should be addressed as part of a comprehensive planning process. In practice, there will always be a need for evaluations of individual projects, policies, etc (Kallis et al. 2004).

Formulation of plans and evaluation of projects is a central issue to the implementation of the Water Framework Directive (WFD). As demonstrated in ADVISOR through the critical review of five *ex-post* evaluation processes, past experience raises questions on whether and how the pursuit of integration and participation may be achieved in "real life" decision processes.

Videira et al. (2003) critically compared the five decision-making processes examined in ADVISOR (see del Moral et al. 2002; Hatzilacou et al. 2002; Hill et al. 2002; van Leeuwen et al. 2002; Videira et al. 2002) with respect to different elements of the participatory processes' design, implementation and assessment. This review highlighted several deficiencies at different stages of the decision-making processes, as summarized below.

Participatory Process Design

- **General Objectives of the Participatory Processes**
  "ADVISOR's WP1 case studies have contemplated a diversity of objectives underlying the development of each of the participatory processes that have been reviewed. In the Portuguese, Greek and Dutch case studies the participatory process was designed for a specific project assessment where a formal EIA process took place. However, the Dutch case study was more integrative in the sense that the process was planned to provide participatory input to the project design from an early stage.
  In the Spanish case participation was driven for consultation purposes. There was not a formal EIA process. The National Water Plan was subjected to a public review period but consultation privileged the National Water Council.
  In the Scottish case study the objective of the consultation focused on the review by the public of the designation process and the measures and actions to be taken".
- **Level of participation impact**
  "With the exception of the Dutch case study, participation has failed to promote higher levels of public impact in the decision-making processes that have been reviewed. In some cases, decision and policy makers have not planned higher levels of public impact[11]*, settling for public information from the beginning. In other situations, the formal consultation processes failed to constitute, in practice, real opportunities for the public to express their opinion about the projects.*
  The Greek case study is a good example of a formal consultation process that ended up as a simple informative process because "the timing was essentially prohibitive for any intervention since the work had already been assigned to a contractor and the constructions had begun".
  The Dutch case study can be regarded as the one that has progressed more towards the stages of Involvement/Collaboration in the public impact spectrum".
- **Types of interested parties**
  "The main actors involved were predominantly the central governments (defending the development of a specific project) and the NGO's (claiming for environment concerns). These parties were usually on opponent sides (Portuguese, Spanish and Greek cases) with the exception of the Scottish case. Apart from the Grensmaas project, the local communities affected by the projects were excluded from active roles in the decision-making process.
  In some cases, it was observed that seemingly relevant stakeholders also didn't participate, e.g. the farmers in the Alqueva case study. Among other factors, there could be indications that the formal consultation methods (public hearings and written statements) were not the most adequate to actively involve "non-technical" people.
  The professional groups had an important role in the Spanish (mainly the irrigators' community) and Greek (mainly the private water companies) case studies.
  The EU played also an important role in the Mediterranean case studies, financing the Projects and in the Scottish case forcing the Scottish Government to designate the river Ythan and estuary as a Nitrate Vulnerable Zone".
- **Timing for participation**
  "The public was essentially involved in the review phase of the evaluation processes, only during the formal consultation period, when the discussions are focused on pre-determined options (if any) and where usually the

---

[11] The spectrum of participatory impact that was considered ranged from information to consultation to involvement to collaboration to empowerment and self determination.

political decision of the project consent is already taken. There was no a priori discussion of the "real problem" and alternatives.

The Evinos case study is paradigmatic in this respect, as the consultation process took place in a stage where the work had already been assigned to a contractor and the constructions had already begun. In the Portuguese case there was an attempt to include participation in the assessment of the project trough Expert Panels; however, at that stage of the process it was very difficult to evaluate other alternatives rather than a dam in the Alqueva.

Only in the Dutch case there seems to have been a thorough problem discussion among the public. We can conclude that a bottom-up discussion of problems is a better arena for "real" public participation than traditional top-down approaches. Public participation must start at the early stages in the evaluation process. This fact is crucial for a truly integrated evaluation process, where affected parties may feel to have contributed to policy solutions and thereby be more committed toward decisions.

Implementation of the participatory processes

"The techniques used to involve the public where mainly public hearings and public comments that were used in the formal consultation period in the review of projects or plans. These methods of formal public consultation were not adequate to involve the stakeholders and local people that don't have a technical/scientific background to discuss the specific impacts of a certain project, whose opinion could be relevant in the identification of problems and alternatives at an early stage in the process, like in the Grensmaas case study.

In the Spanish and Portuguese cases there were also used Advisory Committees, which have played in fact an important role in the evaluation process. The National Water Council in Spain was the main body for the involvement of the public in the decision-making process. CAIA (Alqueva project's accompanying commission), in the Portuguese case study was only related to the post-decision and monitoring phase and proved to be effective in assuring the participation of different stakeholders in the implementation of measures to minimize the environmental impacts of the Alqueva dam".

Assessment of participatory processes' results

"Only in the Dutch case study has the public participation process had a strong impact in the decision-making process. The Alqueva participatory process influenced the measures to be taken especially in the post-decision phase, but in the remaining cases it was shown that the participatory process didn't have an influence in the overall outcomes of the process (decisions, influence on decisions and active design of alternatives, plans, programmes or measures).

With respect to social goals, again, the Dutch case study gave some indications of how the incorporation of public values early in the decision making process can improve the quality of decisions. Neverthless, in the remaining case studies sometimes the participatory process was a way to increment the conflict among competing interests with a confrontation between the central authorities and the environmental NGO's. The Spanish case study is a good example of how implicitly excluding interests may intensify the conflicts among competing parties in the environmental problem at stake".

In face of the above (and considering also the patterns outlined in Chapter 3, the lessons learned from past experiences of project evaluation processes have revealed five main types of methodological and procedural deficiencies:

- **Lack of explicit procedures**
  Other than the Environmental Impact Assessment process, the evaluation of the examined projects and plans did not follow a formal, tractable procedure. There was no explicit, regulated procedure defining the steps that had to be followed for the formulation of a plan or the authorisation of a project. Scientific or economic evaluations were constrained to "rubber-stamping" pre-taken decisions or simply settling details of smaller significance (such as the reservoir's design in the case of Evinos).
- **Inappropriate treatment of scientific uncertainty and complexity**
  There was a lack of reliable scientific data as a basis for sound decisions. Complex information was treated in a subjective way, gaps in knowledge and uncertainties concealed under deterministic statements or calls for further research. For example, in the Ythan case the scientific understanding of the agro-ecosystem nutrient cycle was far from complete but once recommended for designation, the eutrophication of the Ythan estuary became framed as a problem related to agricultural nitrate pollution, despite high scientific uncertainty and different opinions between different government agencies.

- **Ineffective participatory tools and procedural platforms**
The public and affected stakeholders had limited access to information and decisions. Their opinion was asked in ad hoc events (hearings) at a late stage and after main decisions had been framed. There was no explicit incorporation of the results of these hearings in the final decisions. Public input was constrained only to the discussion of the environmental impacts and not the overall rationale of the projects.
- **Neglect of multiple values and view points**
Decisions were based on single, given and non-debated goals. For example, securing more water for economic growth was the unquestionable goal in the Alqueva, Evinos and Spanish NHP cases. In relation there was a limited consideration of alternative dimensions and selection criteria. Technical feasibility and one-off construction costs were the main factors taken into account in the Alqueva, Evinos and Spanish NHP cases. Environmental and social impacts were considered only in a reactive and remedial fashion through the EIA process, after key decisions were made. There was little concern for alternative viewpoints and alternative solutions. For example, the NHP of Spain did not consider in detail water demand management interventions as an alternative to new dams and transfers. In the Alqueva case, different designs of the dam were considered, but not alternatives to the dam that could also fulfill the goal of rural development in the Alentejo region. In fact, it seems that the decision for the specific projects preceded the formulation of goals and plans. The Alqueva and Evinos dams for example, were referred to many times in past decades as strategic reservoirs that had to be constructed. Plans were made to justify their construction, not to really compare them with alternatives.
- **Conflicts leading to ineffective decisions**
Reactions from marginalized communities and stakeholders groups abounded in all five cases and impeded the implementation of the projects, rising also costs. In the Greensmaas case reactions from local inhabitants has blocked progress with the development of the project. Farmers have opposed intensively the designation of Ythan as a nitrates vulnerable zone. Demonstrations against the Spanish NHP in Spain and Brussels contributed to the repeal of the plan by the new government. In the Spanish NHP, it was documented that demand management could save an equal amount of water to this secured by the waterworks, at a significantly lower cost and with significantly lower environmental and social impacts. In the Evinos case, the additional quantity of water secured by the dam proved unnecessary after the end of the drought that prompted its construction. In Ythan, there are concerns that the expensive controls on farms in the designated zone will have only limited impact on the problem of euthrophication.

Whereas guidance has been provided for participatory processes in the context of implementation of the WFD (EC, 2002), the deficiencies presented above indicate that further guidance is still needed on how to link them with actual decision-making and especially evaluations in the planning process. In ADVISOR it is defended that this linkage can be achieved by Integrated Deliberative Decision Processes (IDDPs) and accompanying tools.

An IDDP is one (Kallis et al. 2004):

- In which decision is the outcome of an explicit process of deliberation (assisted by relevant tools) with broad and inclusive participation of stakeholders and the public;
- Based on social robust information and knowledge;
- That takes explicitly into account in its design and conduct the presence of multiple values;
- Embedded in an inclusive and discursive governance framework (institutional, regulatory, and social).

As depicted in Table 4.1, some of the strengths foreseen for an IDDP are related with the potential that these processes offer in terms of tackling the deficiencies that past experiences have exposed.

Table 4.1 Potential benefits of IDDPs

Past Experiences revealed…	IDDPs are expected to…
Lack of explicit procedures	Provide a stepwise procedure for integrated and participatory decision processes; The explicit reference to integration emphasises the need to see the deliberative process not alone, but in relation to the surrounding governance framework and in coordination with the management of information and the handling of multiple values and value systems. An IDDP is not a one-off process or deliberative event, but a longer process consisting of several activities and events, and more importantly, embedded in the surrounding social and policy context.
Inappropriate treatment of scientific uncertainty and complexity	Rely on socially robust knowledge, i.e. knowledge that is fit for the purpose of the process, which cannot be described in limited terms as featuring scientific quality but must be fit for the context and relevant to the concerned social actors; Rest on a "post-normal" scientific paradigm in which an "extended peer community" becomes effective participant in the use of science for decisions (Funtowicz and Ravetz, 1990). Scientific quality assurance depends on an open dialogue among all those affected, quality understood as a contextual property of scientific information. It may include formal or informal procedures, including definition of a set of quality criteria in quality assurance protocols. The "extended community" through appropriate process can then contribute to the co-production of socially robust knowledge.
Neglect of multiple values and view points	Recognize the multiple values present in any decision. This distinguishes it from standard economic valuations which are based on the use of only one type of values (monetary). A deliberative decision process is designed instead, exactly in order to reveal value differences, make the participants aware on such differences, and paving some shared ground for decision-making.
Ineffective participatory tools and procedural platforms Conflicts leading to ineffective decisions	Combine deliberative processes with decision-support systems or formal evaluation and planning methods. Pluralism in the use of tools is an advantage, as is pluralism within the application of each tool; Define clear rules beforehand, in accordance to an appropriate regulatory framework; Account for representation issues; Open up the debate from earlier stages of a decision-making process Aim for higher participatory impact levels; Explicitly account for an evaluation of the participatory process itself.

4.2. Integrated Deliberative Decision Processes Framework

The essence of a deliberative process is that decision (or at least, reaching a point upon which decisions can subsequently be taken) is the result of interaction through dialogue between a broad range of participants with respect for their different values and viewpoints. The framework for IDDPs proposed in ADVISOR is presented in Figure 4.1, which represents a generic process which can be adapted to national, regional and/or local circumstances and to the specific needs of a given decision process. Political, organizational, cultural, hydrological and other environmental conditions should be taken into account when designing each deliberative process.

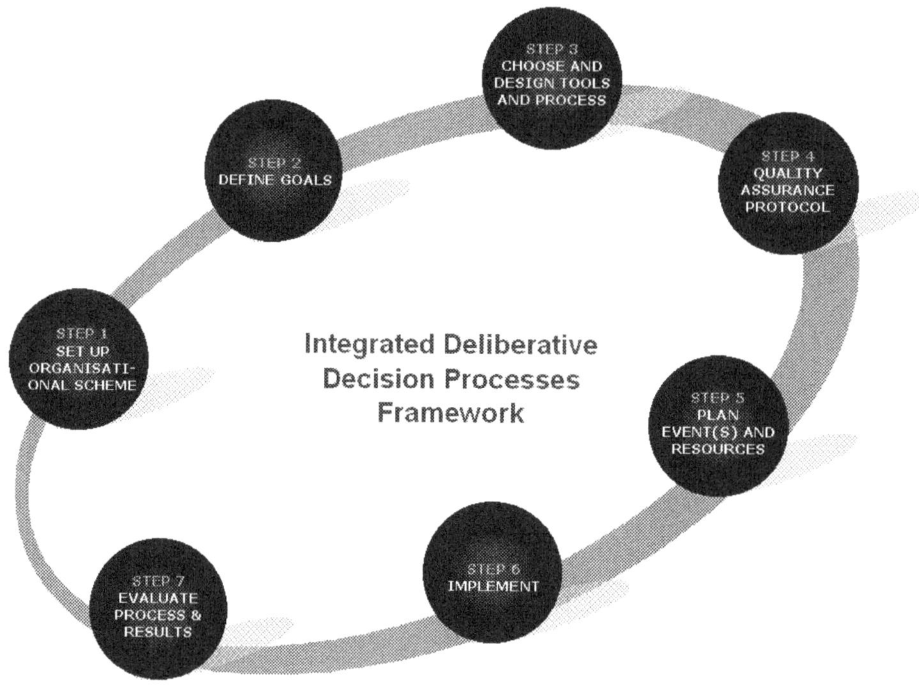

Figure 4.1 An IDDP framework for water resource planning and evaluation

The proposed IDDP framework includes seven basic steps, which are briefly described below. A more detailed description of the framework is presented in the ADVISOR Guidance Document on IDDP (Kallis et al. 2004) (Available from http://ecoman.dcea.fct.unl.pt/projects/advisor).

STEP 1 - Setting up of the organisational scheme

An institutional actor should undertake the responsibility for the whole process and build up a partnership with main stakeholder and citizen groups. A Steering Group of representatives, assisted by professional groups where relevant, should then cater for the planning and implementation of the IDDP.

STEP 2 - Definition of decision and deliberation goals

The issues and questions addressed (i.e. the decision) and the expectations from deliberation should be debated by the partnership and decided at an early stage of the process. Deliberation goals might stem from exchanging information or educating the participants to resolving conflicts of agreeing in common plans and projects.

STEP 3 - Designing deliberative tools and participation process

There exist several procedures, methods and tools to support deliberative decision-making. For any method, one may distinguish between the **procedural platforms** upon which deliberation takes place (e.g. focus groups, workshops, panels, hearings, etc) and the **tools** that help to facilitate the deliberation and the making of the decision (e.g. facilitation procedures, conflict resolution oriented facilitation, scenarios, visions, models, multi-criteria matrices, etc).

Tools should then be selected to fulfil the defined deliberation and decision goals. One or more tools may be used and combined, depending on purposes.

STEP 4 - Establishment of an information Quality Assurance Protocol (QAP)

Clear guidelines and rules, formalised in a protocol, should be used to guide the management of scientific information for the deliberative decision processes.

STEP 5 - Planning of event(s)

The process and the individual deliberative decision events (workshops, group meetings, etc) should be planned carefully well in ahead. Sequential tasks include: preparation of a scoping document with title and expected outcomes of the process; preparation of a detailed work plan with an accompanying time table presenting activities to be carried out, events to be held and participants to be involved; formulation of a detailed budget for the process indicating allocation per category of costs and sources of funding; preparation of specific work plans for the deliberative events to be carried out.

STEP 6 - Implementation of event(s)

This will depend on the tools(s) used. Professional facilitation is crucial for deliberative decision meetings.

STEP 7 - Evaluation of the process and its results

Results of the process should be evaluated both in terms of the quality of the procedure and in terms of contribution to the decision goals. Evaluation should be performed by the organisers together with the partnership of the stakeholders involved and the broader community. An IDDP will be successful only if it really matters. Mechanisms should be established for embedding the outcomes of deliberative processes in real decision-making. An IDDP is conditioned by the constraints set by the surrounding social, institutional and power relationships. Still, reciprocally deliberative processes can build up an impetus for changes in the broader social context.

The remaining chapters of this report illustrate the implementation of some of the IDDP steps according to the results obtained in the ADVISOR project.

Chapter 5 presents a protocol for Quality Assurance by extended Peer Review, which was applied to a case study in Portugal- the Mediated Modelling process in the Baixo Guadiana river basin.

One of the tools tested in ADVISOR was this of a combined cost benefit / cost effectiveness analysis framework methodology to evaluate flood protection projects in which the incorporation of stakeholder participation was proposed is presented in Chapter 6. The results of the case study presented in Chapter 7 support the approach to evaluation which defends that economic values are only a subset of the several types of values that drive human behaviour (in line with IDDP principles). This study extended the conventional Contingent Valuation Method survey to include ethical variables, social psychology drivers and indicators of environmental awareness.

The ADVISOR project has also tested three deliberative techniques: visioning/scenario workshops (See Chapter 8), mediated modelling (see Chapter 9) and social multi-criteria evaluation (see Chapter 10). Together with deliberative monetary valuation and citizens' juries/consensus conferences, these provide a basic (though not exhaustive) toolkit that can serve basic water planning and decision needs. Broader participatory processes such as public hearings, opinion polls, referendums, etc should complement stakeholder or random group deliberation processes.

REFERENCES

Del Moral, L., Pedregal, B., Calvo, M. and Paneque, P. (2002) River Ebro Interbasin Water Transfer, Work Package 1 of project ADVISOR, Universidad de Sevilla.

European Commission (2002) *Guidance on public participation in relation to the water framework directive. Active involvement, Consultation and Public access to information,* Directorate for Environment, Brussels.

Funtowicz, S. & Ravetz, J. (1990) *Uncertainty and Quality in Science for Policy.* Dordrecht: Kluwer Academic Press.

Hatzilacou, D., Kallis, Giorgos and Coccossis, H. (2002) The River Evinos Reservoir, Work Package 1 of project ADVISOR. University of the Aegean.

Hill, G., Urama, K., Spash, C. and Wynn, G. (2002) The Designation of the River Ythan and Estuary as a Nitrate Vulnerable Zone, Work Package 1 of project ADVISOR, The Macaulay Institute.

Kallis, G., Videira, N., Antunes, P., Santos, R., Bruinsma, F., Coccossis, H., Corral Quintana, S., Guimarães Pereira, Â., del Moral. L., Hatzilacou, D., Lobo, G., Paneque, P., Pedregal Mateos, B., Spash, C., Urama, K., Vreeker, R. (2004) *Integrated Deliberative Decision Processes for Water Resources Planning and Evaluation,* Guidance Document prepared for the ADVISOR Project.

Van Leeuwen, E., Dalhuisen, J., Vreeker, R. and Nijkamp, P. (2002) The Grensmaas Project, Work Package 1 of project ADVISOR, Vrije Universiteit Amsterdam.

Videira, N., Lobo, G. Antunes, P., Santos, R. and Guimarães Pereira, Â. (2002) Alqueva Multipurpose Project, Work Package 1 of project ADVISOR, New University of Lisbon.

Videira, N., Lobo, G. Antunes, P. and Santos, R. (2003) Public Participation in Integrated River Basin Governance, Work Package 2 of project ADVISOR, New University of Lisbon.

Chapter 5

Knowledge Assessment: Protocol for Quality Assurance by Extended Peer Review

Serafin Corral Quintana[1] and Ângela Guimarães Pereira[2]

[1] *University of La Laguna, Spain*

[2] *EC - JRC – IPSC. Ispra, Italy*

This chapter deals with the assessment of quality of the information used in Integrated Water Planning Processes. Firstly, the quality of information used in river basin assessment studies is analysed for the 5 case studies available for ADVISOR (Del Moral et al. 2002; Hatzilacou et al. 2002; Hill et al. 2002; Van Leeuwen et al. 2002; Videira et al. 2002). Quality categories are used to produce a diagnosis of a purposeful developed model of information flows, characteristic of water assessment studies. The scheme comprises criteria to analyse information produced and communicated and assessment activities undertaken. Subsequently, both a normative and a descriptive use of the quality categories is proposed in a form of a protocol for quality assurance, illustrated with an application to a case study in Portugal: The Mediated Modelling process in the Baixo Guadiana, suggesting its usefulness in the planning activities envisaged under the WFD.

5.1. INFORMATION FLOWS MODEL

Three main streams of analysis of information inputs in the process of evaluation of river basin projects have been considered; information used for:

- framing → understanding & setting context
- assessment activities
- communication with stakeholders → public participation

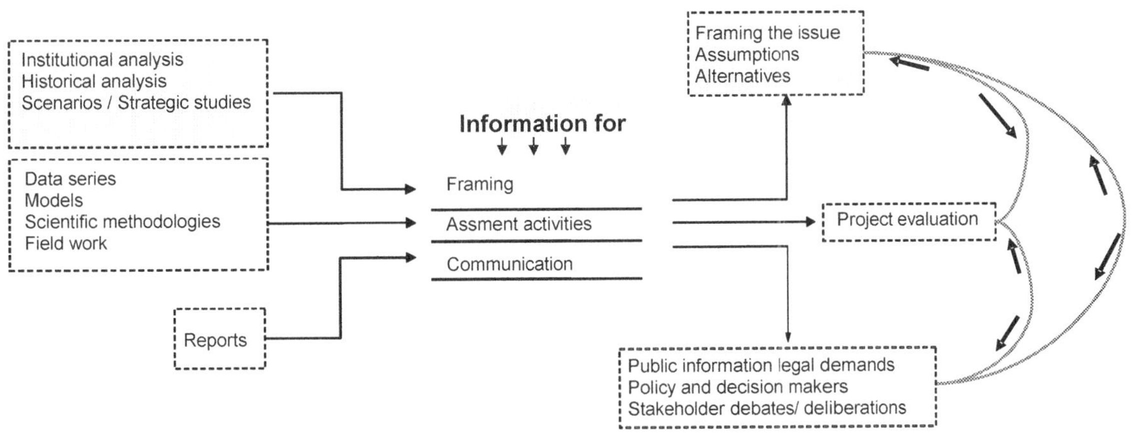

Figure 5.1 Possible information streams in river basin assessment

Figure 5.1 illustrates the streams of information that flow into the process and possible interactions among the three, that is new flows of information emerging from the creation of knowledge during the several stages of the process may feed other phases (either subsequently or iteratively). Information requirements for the assessment activities are very much determined from the outcomes of the problem formulation phase and that in turn could lead to an iterative process of refining problem frames and definitions (in an ideal situation). Also, ideally, it is within the assessment activities that the 'publics' get consulted and yet through the very consultation process using some scope activity, the information requirements to do the assessment can be set. Hence, the information flows of the whole process dynamics is essential to reason upon the other ADVISOR's tetrahedron vertexes: context, assessment and participation, and indeed the review of the studies carried out for the information vertex will address the other vertexes of the tetrahedron.

Eventually, the question we will try to answer is what was the information used in the decision making process and what is the context of its application, those determining largely the information quality assessment of the 5 case studies. The analysis of the information flows is done in the framework of quality assessment. The quality criteria relate for instance to the:

- Types and sources of information, process phase and timing of application
- Pedigree of information (including fitness for purpose, legitimacy, uncertainty, etc.)
- Transparency of reported information (to stakeholders, decision/policy makers, etc).

5.2. QUALITY DIAGNOSIS: THE CASE STUDIES

5.2.1. Case study UK: River Ythan and Estuary as a Nitrate Vulnerable Zone

This section is based on the case study review report carried out by the UK team, Hill et al. (2002). The information used or produced during the whole process falls in the following categories:

- Peer review scientific publication
- Non-peer review scientific publication
- Unpublished but accessible environmental data
- Information for public consultation
- Specialist media, e.g. interest group publications, official reports of parliamentary debate
- Press and general media articles.

The audiences and the levels of the process at which information was deployed are varied. The first three categories of information were mostly used for policy purposes, whilst other categories of information could be accessible to other audiences. Peer reviewed information is clearly the most reliable but not necessarily the most accessible for all those concerned and therefore it cannot be concluded immediately that the information available was fit for the purpose.

Information used in this case study seems to have been highly controversial, surrounded by a great deal of uncertainties and evidence provided by the scientific community was not shared at policy level. The evolution of the process in time did bring some further information but lacks of information have remained and eventually decisions were made invoking the Precautionary Principle.

In this section the analysis of the quality of information is summarised for the phases of the process across which this analysis has been carried out. Tables 5.1 and 5.2 propose for the problem framing and assessment phases of the process an appreciation of the quality of information used and activities undertaken, whereas Table 5.3 relates to information used for communication purposes.

Table 5.1 Quality of information used in the assessment

Fitness for purpose	
Adequacy & relevance	Information used seems to be specific for the problem and relevant
Accuracy	Uncertainties are acknowledged.
Completeness	Information was not comprehensive at all steps of the process.
Applicability	
Access & Availability	Available information was accessible to those concerned with the assessment activities.
Intelligibility	Information was mostly intelligible for expert usage.
Reliability - CONTROL	
Sources of Information (where)	Literature; research project reports; different types of publication; specialist media and press; consultation processes
Sources of information (methods used)	Not clear from the reports what methods were used, but possibly field analysis. No formal methods are stated to estimate economical costs.
Sources of information (generators)	Environmental agency, scientists
Verification & Colleague consensus	Some information has been peer reviewed.
Reliability - CONFIDENCE	
Extended peer acceptance	Not enough information to do an overall judgement. However, in the beginning of the process there was controversy to whether information provided could be used as evidence for NVZ designation
Legitimacy	The main sources of information are from the scientific community.

Table 5.2 Quality of activities undertaken

Fitness for purpose	
Adequacy & applicability	No formal assessment process; no formal methods were used according to the reports.
Adaptability	Not known.
Transformation/encoding	Not known.
Transparency	
Documentation	No overall formal assessment process; hence difficult to respond to this issue. It can be thought that peer reviewed sources are documented.
Sources	No overall formal assessment process; hence difficult to respond to this issue.
Legitimacy	
Colleague consensus	As there was no formal assessment processes and methodologies, this item is considered quite low.

Table 5.3 Quality of information used for communication purposes

Intelligibility	
Compliance with audience	It seems that information for public consultation was largely tailored to the type of consultees. However, information was mostly technical and probably not widely intelligible for all.
Accessibility	It seems that information for public consultation was largely accessible to the type of consultees. However, information was mostly technical and probably not widely accessible for all.
Transparency	
Documentation	Reasoning behind the conclusions drawn by policy makers were not made public in due time.
Communication of uncertainties	
Statement of uncertainties	Uncertainties stated. Opportunity to disagree with scientific evidence has been granted during consultation
Implications for decision stated	The Precautionary principle was adopted as justification of decisions.
Pedigree of results	Not known, but in principle that was not available.

5.2.2. Case Study Spain: River Ebro Interbasin Water Transfer

This section is based on the case study review report carried out by the Spanish team, Del Moral et al. (2002). The information used or produced during the whole process falls in the following categories:

- Historical data sets
- Modelling and simulation data (some models peer reviewed, others not)
- Economical and social analysis
- Field work
- Press

The first four categories of information were mostly used for policy purposes, whilst the press deployed information accessible to other audiences. Peer reviewed information is clearly the most reliable but not necessarily the most accessible for all those concerned and therefore it cannot be concluded immediately that the information available was fit for the purpose.

The information used seems to be in many cases inadequate and inaccurate and so not fitting the purpose. The information used is mainly technical in nature and it was very contested by peers.

There were many controversies about the problem framing and generation of alternatives. In fact, the assumptions made to simply design the current PHN are not shared among an extended community of peers.

Uncertainties are poorly managed within the assessment studies. Either they are not acknowledged or they are overlooked. Decisions were taken anyway, despite uncertainties and gaps of knowledge.

There has not been a tailoring process of the information to make it available to wider audiences, yet the public hearings and consultation that have taken place have produced reports with main concerns. The way in which those were incorporated in the process remains unclear, but it is believe that in informal ways the process has been influenced by those outcomes.

In this section the analysis of the quality of information is summarised for the phases of the process across which this analysis is performed. Tables 5.4 and 5.5 propose for the problem framing and assessment phases of the process an appreciation of the quality of information used and activities undertaken, whereas Table 5.6 relates to information used for communication purposes.

Table 5.4 Quality of information used in the assessment

Fitness for purpose	
Adequacy & relevance	Information used seems to be relevant for the problem but sometimes not adequate.
Accuracy	There are many inaccuracies in the information produced. Uncertainties are not always stated. There are many gaps of information and inappropriateness of some model results jeopardises accuracy of subsequent recommendations and statements.
Completeness	Information was not comprehensive. There were many gaps in fundamental issues.
Applicability	
Access & Availability	There is little information about this as to whether so much reported gaps are due to accessibility and availability problems or to the choice of not considering the issues.
Intelligibility	Information was intelligible mostly for experts. There is also a great deal of propaganda material.
Reliability - CONTROL	
Sources of Information (where)	There are problems with the statement of the sources of information.
Sources of information (methods used)	Analysis of historical data; field measurements; models, public hearings.
Sources of information (generators)	Scientific experts and stakeholders.
Verification & Colleague consensus	A great deal of controversy related to data used. There is no consensus in many of the themes. Information has been subject to peer scrutiny, but the outcomes were in many cases largely ignored.
Reliability - CONFIDENCE	
Extended peer acceptance	Information has been scrutinised in public consultation, and reports were done.
Legitimacy	Legitimacy is highly questioned. Many controversies, little confidence.

Table 5.5 Quality of activities undertaken

Fitness for purpose	
Adequacy & applicability	Activities undertaken are seen as many times inadequate, models used, assumptions made, framings, etc. There is hardly any discussion about the applicability conditions of models or methods proposed through the study.
Adaptability	Not enough information to answer this, but there were questions about to whether some sophisticated modelling was actually just a legitimating mechanism of decisions taken.
Transformation/encoding	Not enough information to answer this.
Transparency	
Justification	Models usage is not clearly justified and documented.
Sources	Sources are poorly documented. Frequently the authors of subsidiary studies are not referenced.
Legitimacy	
Colleague consensus	High controversies on methods and models used, problem framing and assumptions made.

Table 5.6 Quality of information used for communication purposes

Intelligibility	
Compliance with audience	There is no information tailored for non-experts. Information ranges from expert and technical information to propaganda material.
Accessibility	Information was accessible, upon request but also through the INTERNET.
Transparency	
Documentation	Sources are poorly documented. Frequently the authors of subsidiary studies are not referenced.
Communication of uncertainties	
Statement of uncertainties	Uncertainties are poorly managed. Uncertainties related to available information (flow records, demand evaluations, etc.) or to key socio-economic processes (e.g. future markets for agrarian commodities) are usually ignored or underestimated. In many cases uncertainty is not an issue. In some cases further studies are recommended but decisions are not stopped if they do not exist.
Implications for decision stated	No implications for decision making. Decisions are taken anyway. Precautionary principle largely ignored.
Pedigree of results	Not known, but in principle this was not available.

5.2.3. Case study The Netherlands: The Grensmaas project

This section is based on the case study review report carried out by the Dutch team, Van Leeuwen et al. (2002). The information used or produced during the whole process falls in the following categories:

- Historical data sets
- Modelling and simulation data (some models peer reviewed, others not)
- Field work
- Press and Internet material

The first three categories of information were mostly used for policy purposes, whilst the press deployed information accessible to other audiences. Peer reviewed information is clearly the most reliable but not necessarily the most accessible for all those concerned and therefore it cannot be concluded immediately that the information available was fit for the purpose.

Information used in this case study seems to be quite fit for the purpose, despite identified gaps and uncertainties. The information used was mostly of technical nature but information from stakeholders seem to have been incorporated at many steps of the appraisal and assessment processes (scoping, review, etc. seemed to have been inclusionary).

The EIA process seem to be in the right path of inclusionary processes, however the way uncertainties seem to have been dealt with is not satisfactory; it is not clear for instance as to whether enlarged participation could help dealing with framing uncertainties and develop more insight.

In this section the analysis of the quality of information is summarised for the phases of the process across which this analysis is performed. Tables 5.7 and 5.8 propose for the problem framing and assessment phases of the process an appreciation of the quality of information used and activities undertaken, whereas Table 5.9 relates to information used for communication purposes.

Table 5.7 Quality of information used in the assessment

Fitness for purpose	
Adequacy & relevance	Information used seems to be specific for the problem and relevant. Some information is said to be inadequate, especially coming from simulation models.
Accuracy	Uncertainties are stated but gaps of information and inappropriateness of some model results may jeopardise accuracy of subsequent recommendations and statements is poor.
Completeness	Information was quite comprehensive but further studies are recommended in many categories.
Applicability	
Access & Availability	Available information was accessible to those concerned with the assessment activities.
Intelligibility	Information was intelligible for experts.
Reliability - CONTROL	
Sources of Information (where)	Literature; research project reports; different types of publication. Sources are documented.
Sources of information (methods used)	Analysis of historical data; field measurements; models, public hearings.
Sources of information (generators)	Scientific experts and stakeholders.
Verification & Colleague consensus	Controversy is not related to data or models used.
Reliability - CONFIDENCE	
Extended peer acceptance	Information has been scrutinised in public hearings.
Legitimacy	There seems to be no controversies on data used, but there are many uncertainties and gaps.

Table 5.8 Quality of activities undertaken

Fitness for purpose	
Adequacy & applicability	The EIA study allowed a number of issues to be covered. However, it is recognised that some themes have not been analysed that are seen as important (financial analysis, e.g.) In general, models and other activities undertaken seem to be adequate and applicable (clear exception to the scale models).
Adaptability	Some models seemed to have been purposefully developed for the river Meuse.
Transformation/encoding	Not known.
Transparency	
Justification	Models usage is not clearly justified and documented, but when describing model's outcomes there is an implicit justification for its usage.
Sources	Sources of models used are documented.
Legitimacy	
Colleague consensus	It is not clear whether those models have been peer reviewed and are consensual among peers. Controversies don't seem however to be related to data or models.

Table 5.9 Quality of information used for communication purposes

Intelligibility	
Compliance with audience	All the reports are available for policy makers and the public and there is information tailored for different audiences.
Accessibility	Information is accessible, through several mechanisms. There is a possibility to order them or the more integrated reports can be downloaded from the project website, from the website of the ministry of transport and public works or from the website of the province of Limburg. Also newsletters, leaflets and meetings contributed to the provision of information.
Transparency	
Documentation	Sources of information are stated and so are methodologies
Communication of uncertainties	
Statement of uncertainties	Uncertainties are stated. In some cases gaps of information are recognised and further studies are suggested
Implications for decision stated	Uncertainties are managed through sensitivity analysis and the study states that uncertainties will not influence the decision. Precautionary principle largely ignored.
Pedigree of results	Not known, but in principle this was not available.

5.2.4. Case study Portugal: The Alqueva Multi-purpose project

This section is based on the case study review report carried out by the Portuguese team, Videira et al. (2002). The information used or produced during the whole process falls in the following categories:

- Fieldwork;
- Modelling studies and sediment balance;
- Historical data sets;
- Spatial database analysis
- Press notes, radio and newspapers
- Propaganda material about the Alqueva dam, such as leaflets, info-points, web sites and videos.

The first four categories of information were mostly used for policy purposes, whilst all propaganda material was deployed to inform a wider audience. Peer reviewed information and methodologies is clearly the most reliable but not necessarily the most accessible for all those concerned and therefore it cannot be concluded immediately that the information available was fit for the purpose.

Information used in this case study does not seem to have been highly controversial; what is controversial is the assumption that the Alqueva dam project is the only solution for regional development and this latter justification be used as the basis to construct it. Still, there are a great deal of uncertainties, information gaps, etc. that have led to arbitrary assumptions, statements and eventually recommendations. The information used was mostly of technical nature but the scoping of the study was peer reviewed by an external panel to the team that produced it, which included not only experts but also members of relevant NGOs. Apparently, the review of the study has also produced some statements that could be incorporated in the study in the form of mitigation measures. That indicates, that although the study was a legitimating procedure to go ahead with decisions taken in grounds other than technical, there was still room for some review.

In this section the analysis of the quality of information is summarised for the phases of the process across which this analysis is performed. Tables 5.10 and 5.11 propose for the problem framing and assessment phases of the process an appreciation of the quality of information used and activities undertaken, whereas Table 5.12 relates to information used for communication purposes.

Table 5.10 Quality of information used in the assessment

Fitness for purpose	
Adequacy & relevance	Information used seems to be specific for the problem and relevant.
Accuracy	Uncertainties are many times ignored and hence accuracy of subsequent recommendations and statements is poor. The greatest uncertainty is related to the quality of the water that could question the whole project construction if the irrigation option prevails as the main objective.
Completeness	Information was not comprehensive at all steps of the process. Many gaps.
Applicability	
Access & Availability	Available information was accessible to those concerned with the assessment activities.
Intelligibility	Information was mostly intelligible for expert usage.
Reliability - CONTROL	
Sources of Information (where)	Literature; research project reports; different types of publication. Sources are usually documented.
Sources of information (methods used)	Analysis of historical data; field measurements; many simulation models, geographic reasoning models; empirical models;
Sources of information (generators)	Scientific experts
Verification & Colleague consensus	Some information has been peer reviewed. Much controversy about assumptions makes also questioning about the results.
Reliability – CONFIDENCE	
Extended peer acceptance	Several statements coming from the public hearings point out "errors, gaps, or the need for further studies. The main areas addressed were climate, geology and seismic risk, ecological flows and general hydraulic studies, physical desertification, biological systems and proposed mitigation measures and socio–economics."
Legitimacy	The main sources of information are from the scientific community; information available is not generally prone to controversy; it's the lack of it that generates controversy.

Table 5.11 Quality of activities undertaken

Fitness for purpose	
Adequacy & applicability	An IEIS was done along with other studies. The scoping was shared in an extended peer review fashion (included experts and NGOs). The models and other activities undertaken seem to be adequate and applicable, but not all are peer reviewed.
Adaptability	Many models have been adapted to the situation, justified as "used in other situations".
Transformation/encoding	Not known.
Transparency	
Documentation	Models usage poorly justified but usually documented.
Sources	Sources of models used are documented.
Legitimacy	
Colleague consensus	Not so much controversy in models usage.

Table 5.12 Quality of information used for communication purposes

Intelligibility	
Compliance with audience	There is no information that there has been an effort to tailor information to be used in public spheres. Even the non-technical information for public hearings purposes is considered still too technical.
Accessibility	Information was fairly accessible in specific institutions; however, 30 days seems to be too few for an serious consultation process.
Transparency	
Documentation	Conclusions and recommendations are based on the studies done – not sure if the technical summaries provided all information about sources, etc.
Communication of uncertainties	
Statement of uncertainties	Not clear if uncertainties are stated. However, in some cases further studies are suggested.
Implications for decision stated	The process ignored many uncertainties; decisions to go ahead were taken anyway. Precautionary principle largely ignored.
Pedigree of results	Not known, but in principle this was not available.

5.2.5. Case study Greece: The River Evinos Reservoir

This section is based on the case study review report carried out by the Greek team, Hatzilacou et al. (2002). The information used or produced during the whole process falls in the following categories:

- Historical data sets
- Modelling and simulation data (some models peer reviewed, others not)
- Field work
- Press

The first three categories of information were mostly used for policy purposes, whilst the press deployed information accessible to other audiences. Peer reviewed information is clearly the most reliable but not necessarily the most accessible for all those concerned and therefore it cannot be concluded immediately that the information available was fit for the purpose.

Information used in this case study seems to have been highly controversial, surrounded by a great deal of uncertainties, information gaps, etc. that led to arbitrary assumptions, statements and recommendations. The information used was mostly of technical nature and no information from stakeholders seem to have been incorporated at any time in the process (scoping, review, etc. were not inclusionary).

It seems as though that the development of the study was a mere legitimating procedure to go ahead with decisions taken in grounds other than technical.

It can be said that in this process information used, especially in the EIA, is of poor quality: poor fitness, reliability and transparency.

In this section the analysis of the quality of information is summarised for the phases of the process across which this analysis is performed. Tables 5.13 and 5.14 propose for the problem framing and assessment phases of the process an appreciation of the quality of information used and activities undertaken, whereas Table 5.15 relates to information used for communication purposes.

Table 5.13 Quality of information used in the assessment

Fitness for purpose	
Adequacy & relevance	Information used seems to be specific for the problem and relevant, but many times not adequate.
Accuracy	Uncertainties are many times ignored and accuracy of subsequent recommendations and statements is poor.
Completeness	Information was not comprehensive at all steps of the process. Many gaps.
Applicability	
Access & Availability	Available information was accessible to those concerned with the assessment activities.
Intelligibility	Information was mostly intelligible for expert usage.
Reliability - CONTROL	
Sources of Information (where)	Literature; research project reports; different types of publication. Sources are documented.
Sources of information (methods used)	Analysis of historical data; field measurements; models on erosion and hydrogeological assessment (the Bemermhn model).
Sources of information (generators)	Scientific experts
Verification & Colleague consensus	Very little information has been peer reviewed. Much controversy.
Reliability - CONFIDENCE	
Extended peer acceptance	NGOs objections not taken into account. Consultation was a sham.
Legitimacy	The main sources of information are from the scientific community, but not all shared.

Table 5.14 Quality of activities undertaken

Fitness for purpose	
Adequacy & applicability	An EIA was done together with other studies. At least for the EIA some themes are obligatory. The models and other activities undertaken seem to be adequate and applicable.
Adaptability	Not known.
Transformation/encoding	Not known.
Transparency	
Documentation	Models usage poorly justified and documented.
Sources	Sources of models used are documented.
Legitimacy	
Colleague consensus	Controversial use of some models.

Table 5.15 Quality of information used for communication purposes

Intelligibility	
Compliance with audience	There is no information that there has been an effort to tailor information to be used in public spheres.
Accessibility	Information was poorly accessible, not only because it consisted of a study placed in some institution, but also because of the time given for consultation.
Transparency	
Documentation	Conclusions and recommendations are based on the studies done but as the studies were poor, so are the results.
Communication of uncertainties	
Statement of uncertainties	Not clear if uncertainties are stated. However, in some cases further studies are suggested
Implications for decision stated	The process ignored completely uncertainties; the decision to go ahead were taken before the studies were done. Precautionary principle largely ignored.
Pedigree of results	Not known, but in principle this was not available.

5.3. FINDING PATTERNS

Information used in the 5 case studies is of all sorts: quantitative, qualitative data, in several formats and from several sources. The remaining of this section summarises main findings in relation to quality of data used, activities undertaken and style of communication.

5.3.1. Quality of Data Used

Fitness for purpose

- Adequacy and Relevance: in general for all studies the information used seems to be specific to the case study or generic to the issues covered
- Accuracy: all studies report high scientific uncertainties and inconclusive nature of scientific evidence
- Completeness: all case studies seem to be bound by great omissions of information. There are two situations:
  - Omission of data is recognised within the assessment studies → the strategy is usually to call for further studies or use data/results from other studies
  - Omission is NOT recognised within the assessment studies but viewed by reviewers as essential.

Applicability

- Access & Availability: access to available information for assessment purposes does not seem to be a problem in any of the studies.
- Intelligibility: information for assessment purposes seems to be highly specialised, essentially intelligible for peers, in the majority of the cases.

Reliability - CONTROL

- Sources of information (WHERE): Scientific Literature; Research project reports; Institutional publications (governmental and others); Expert/scientific consultancy – very little if any information is coming from social research.

- Sources of information (Methods used): Calculations; Estimations (scenarios, model simulations, etc.); Historical values/series/etc.; Field work
- Sources of information (Generators): Mainly experts and consultants
- Verification & Colleague consensus about sources: In many cases information used and produced is controversial.

Reliability - CONFIDENCE

- Extended peer acceptance: Information in many cases has a low degree of acceptance by the non-scientific communities
- Legitimacy: Information coming from peer reviewed scientific literature seems to be legitimate; reports suggest sometimes that technical studies done for the assessment are not credible.

5.3.2. Quality of Activities Undertaken

This section refers to methodologies, models, etc used in the assessment process.

Fitness for purpose

- Adequacy: some reports suggest that some of the methods or conceptual approaches were NOT adequate: e.g. some of the expressions used are, conceptual errors, overestimations, omissions, underestimations, wrong assumptions, inadequacies…
- Adaptability: some studies have deployed models which may be generic – but it is not clear if they are indeed (e.g. adoption of US EPA models in European cases); some methodologies seem to have been tailored or purpose built for the case studies.
- Transformation/encoding: large range of transformation types – extrapolations; aggregations; etc.

Transparency

- Documentation: usually sources are stated, but less usually though are explanations, justifications or statement of assumptions made in relation to models, methods, etc used in the assessment communicated through documents aimed at non-specialised audiences
- Arbitrariness: see above.

Legitimacy

- Colleague consensus: in many cases the methodologies used were not consensual among peers.

5.3.3. Style of communication

Intelligibility

- Compliance with audience: reports suggest that the studies done are suitable for technical and scientific audiences; in many cases, it is not clear whether the assessment reports are suitable for non-scientific audiences, including policy makers. Some assessment studies (e.g. developed within EIAP) require the availability of a non-technical reports or consultation documents and so, in principle these were the sources compliant with non-expert audiences. Also, many times the press and propaganda material were the sole intelligible sources of information for non-experts, yet it is not surely the desirable or effective means of wider involvement of relevant communities.
- Accessibility: Some assessment studies (e.g. developed within EIAP) require the availability in specific locations. Internet was also used to make reports available. In terms of accessibility it must be noted however, that in many cases the time available for consultation, participation, etc. was not satisfactory.

Transparency

- Documentation: difficult to appreciate this issue with information provided in the reports It seems as though that sources of information were clearly stated, but there is little basis to conclude that models' assumptions, justification, applicability conditions or necessary adaptations, have been available at all, unless full reports were used (and when those contained such types of information).

Communication of uncertainties

- Statement of uncertainties: In some studies acknowledgement of uncertainties (including ignorance) and major gaps of information is done. For some studies uncertainty is not an issue.
- Implications for the decisions stated: This seems to be lacking for the majority of the cases. In the majority of the cases, decisions are made, despite uncertainties (or ignoring them). The Precautionary principle is largely ignored.
- Pedigree of results: there are no such types of information available; however, in some cases other studies or sources are referred to.

In general dealing with uncertainties throughout the studies is treated in the following way:

- Not an issue
- Suggestion for further studies
- Go ahead despite uncertainties: implications for the formulation of the problem and for the assessment phases are not clear
- Precautionary principle seems to be largely ignored (in one case study it is suggested that in-action was adopted as a precautionary measure)

Therefore, since the quality of the information used in the studies will backup decisions, it emerges from here that a different type of control has to be done, to have information that is relevant from the societal point of view and that can only be achieved through extended peer review processes with those concerned.

5.3.4. Information from actors into the process

Rarely there seems to be any type of consultation for the formulation of the problem or the definition of scope of the assessment studies (exception is for instance the Alqueva IEIS, where an external panel was invited to determine the scope of the study). It seems as though that post-reviews are more favoured than ant-reviews.

When the participatory process occurs, it is not clear how recommendations are integrated in the assessment studies and/or decision processes – and actually whether they are at all considered in the process (evaluation and decision). In some cases, however, concerns from public hearings have been incorporated into mitigation measures and when controversy was so high, so, not really public participation, but actual conflict was installed, the whole appraisal process was jeoperdised (example: Spanish and Dutch case studies).

In most cases there seems to be just a one-way information flow: information from the proponents of the projects towards the public. In few case studies there are suggestions that actual consultations took place, the results of which are far from being normative and recommendable (e.g. no formal ways of incorporation of the consultation outcomes, timings prohibitive for serious involvement, etc).

5.4. PROTOCOL FOR KNOWLEDGE QUALITY ASSURANCE of WATER PLANNING and GOVERNANCE PROCESSES

The following protocol (see Annex I) was developed taking into account the results of the previous sections. Although, it is specifically tailored for a case study developed under the ADVISOR project, deploying a specific research methodology, participatory modelling, it contains the necessary ingredients to be used widely in other types of evaluation deploying different planning processes and methodologies.

5.4.1. Structure of the Quality Assurance Protocol

Categories for quality assessment and information pedigree developed by Corral Quintana (2000) will be presented. Some of these categories were tailored to design a Quality Assurance Protocol that was applied in the Portuguese case study of the ADVISOR project to quality assure information, tools used and communication. The quality assurance process is carried out as a combination of an extended peer review12 and a scientific peer review the latter dealing with more technical aspects of the methodology and information.

[12] In environmental and health problems, the maintenance of quality depends on an open dialogue among all those affected. This is what Funtowicz & Ravetz call an *"extended peer community"*, consisting not merely of persons with some sort of institutional accreditation, but rather of all those with a desire and commitment to participate in the resolution of the relevant issue (Funtowicz and Ravetz, 1990). Their contribution being not merely a matter of broader democratic participation, but a contribution to the co-production of knowledge. The procedures for assessing the quality of such policy processes is an "extended peer review", performed by an extended peer community.

Together with the previously defined and applied categories to assess the Quality of the case studies, the QA protocol designed to be implemented in the Portuguese case study included another three new categories not used previously to assess the quality of the 5 ADVISOR case studies. So the QA protocol was divided into the following six areas:

- framing
- social engagement
- research team's role
- research procedures undertaken
- background information (scientific & technical information)
- communication of information.

The number and order of these categories is based on the characteristics of the methodology applied in this case study. The organisation of the mediated modelling process determines the structure of the QA protocol. Hence, the decisions taken by the research team of choosing a mediated modelling approach and holding three participatory workshops with the participants in which the assessment is progressively developed (see Figure 5.2) has determined the structure of the protocol.

Framing and the social engagement categories address the assessment of how the problematique is defined and on this basis which are the procedures to identify who should participate in the workshops. These two categories represent the first task to be carried out by the research team, this explains their initial position in the protocol.

The research team's role is an import aspect to be assessed because the Portuguese research team was, among other tasks, in charge of:

- providing information to the participants to the workshops,
- facilitating and collecting information for such participatory process, and
- translating this input into quantitative results.

The assurance process of the scientific and technical information is placed at the end of the protocol. This type of information is used because the information plays an important role in the assessment when the research team "translates" the model developed together with the participants into figures. On the contrary the research team's role is located at the beginning of the protocol.

5.4.2. Quality Assessment Categories

In this section we present the categories for quality assessment and information pedigree to assess the quality of the information used in the 5 ADVISOR assessment studies. The three activities are assessed using a set of criteria specifically defined to analyse the different attributes that characterise each activity. This scheme is based on the protocol developed by Corral Quintana (2000) (see Annex II).

Framing & social engagement:

When looking into framing, one seeks to understand what was the basis to initiate a process of river basin governance in the chosen terms, i.e. how problems have been formulated taking into account the context in which they developed. Therefore issues such as historical and institutional basis; context description; stakeholders identification (who has initiated the debate); etc. that can provide the justification basis for framing the problem in the socio-economic-political context, have to be looked at. Also, the context dynamics influence on information requirements and use will be analysed. The alternatives chosen to tackle the problem, their origin, or generation processes are also looked at to assess the quality of the problem formulation phase, as well as the assessment activities.

There are now many initiatives, increasing in number and significance, for involving wider circles of people in decision-making and implementation of health and environmental policy-related issues. "In the sorts of issue-driven science relating to the protection of health and the environment, typically facts are uncertain, values in dispute, stakes high, and decisions urgent.

All too often, hard policy decisions have to be made where only scientific inputs are irremediably soft. The requirement for the "sound science" that is frequently invoked as necessary for rational policy decisions may affectively conceal value-loadings that determine research conclusions and policy recommendations" (Funtowicz and Ravetz, 1990). A more relevant and robust guiding principle is quality, understood as a contextual property of scientific information (Funtowicz and Ravetz, 1990; Ravetz, 1996; Funtowicz, 2001).

In environmental and health problems, the maintenance of quality depends on an open dialogue among all those affected. This is what Funtowicz & Ravetz call an "extended peer community", consisting not merely of persons with some sort of institutional accreditation, but rather of all those with a desire and commitment to participate in the resolution of the relevant issue. Their contribution being not merely a matter of broader democratic participation, but a contribution to the co-production of knowledge. The procedures for assessing the quality of such policy processes are an "extended peer review", performed by an extended peer community.

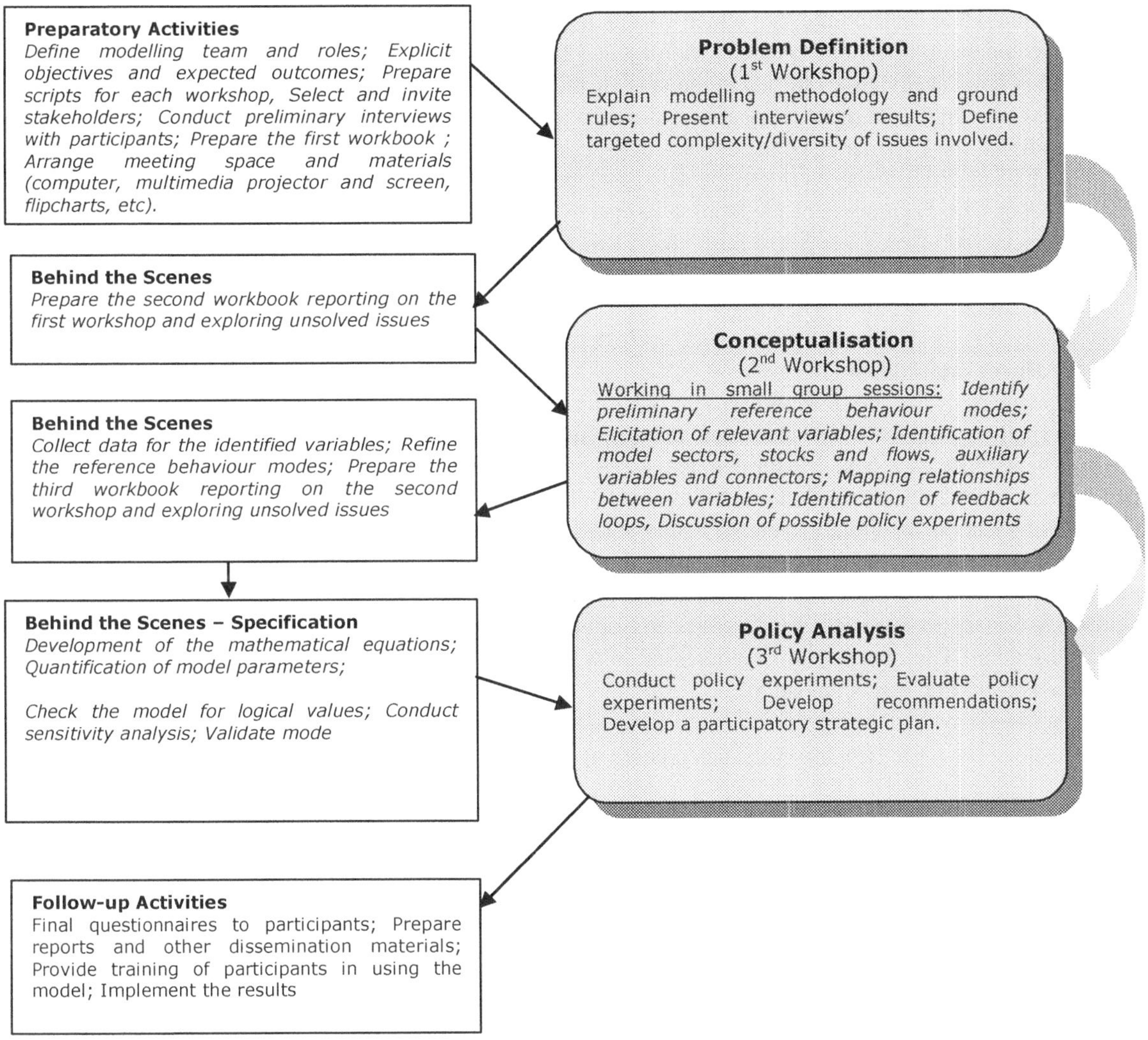

Figure 5.2 Stages of the Mediated Modelling Process in the Baixo Guadiana (Videira et al. 2004)

Research team's role:

Analysts play an important but frequently forgotten role in the problem-solving process that accompanies decision-making processes. They are in charge of transforming the socio-environmental issue at hand into a decisional problem, defining criteria and alternatives that reflect socio-political perceptions about the issue and choosing the evaluation engine to be applied in the assessment process.

Usually the role played by experts during the analysis is considered neutral. This belief should be revisited when dealing with complex issues, where the degree of uncertainty and ignorance, as well as the need for urgent decisions in a context of high stakes and social, economical, environmental and political interests and conflicts are more than evident. In these cases the analysts' experience and their acceptance by social actors will play a crucial role in the process of framing the problem-solving process.

Experts are not intended to replace the decision-maker in solving the problem, but to formulate the peculiarities of the problem in a clear, concise and structured way. On the basis of the objectives expressed in relation to the recognised problematique by social actors, the analyst will examine the existing data in order to develop the hypothesis and criteria to be used during the analytical phase of the decision-making process.

Since it is recognised to experts their capability of translating the decision issue into alternatives and criteria it has also to be recognised the planning process, and therefore its quality, may be influenced by that process of transformation.

Quality Assurance categories are used in this case to assess that process, paying special attention:

- To understand the problematique to be dealt with, and
- To transform information into alternatives, hypothesis and criteria.

Experts also choose the decision-aid tool which will be used to assess the different policy alternatives. It is clear the election of one technique in detriment of other will influence the results of the analysis.

Two categories have been chosen to tackle the role of the analyst: **Competence** and **Legitimacy**. Competence is intended to assess the experience of the analyst in addressing a particular policy issue, in that sense three characteristics have been considered: the Experience of the experts in processing these problems; their Adaptability to deal with new components or characteristics of the issue and the way in which problems are structured.

Legitimacy copes with the process of verification passed by the methodology - either developed or implemented - by the analyst. A process of verification where both, the scientific community and the community at large -those relevant in planning issues: stakeholders and population- are involved. Legitimacy is assessed by means of two categories: Control and Acceptance.

Research procedures undertaken:

The quality assessment of the activities undertaken to evaluate river basin governance projects consist not only of the analysis of information used (including data sets, methodologies) through criteria of pedigree of information, but also the ways in which those that have performed the assessment have dealt with often observed information related problems (such as uncertainty, controversy – different schools of thought – data collection problems, knowledge gaps, generic models application, assumptions, etc.). Quality refers also to the transparency of the assessment procedures and to the robustness of the results that are made available through the studies. Robustness of results has to be assessed for their conclusiveness, their framing given the assumptions made at an earlier stage in the assessment, etc.

In Figure 5.3, we try to illustrate the several steps often seen in environmental assessment studies and possible sources of information. The question to be addressed is what have been the processes of informing the assessment activities of the river basin project evaluation?

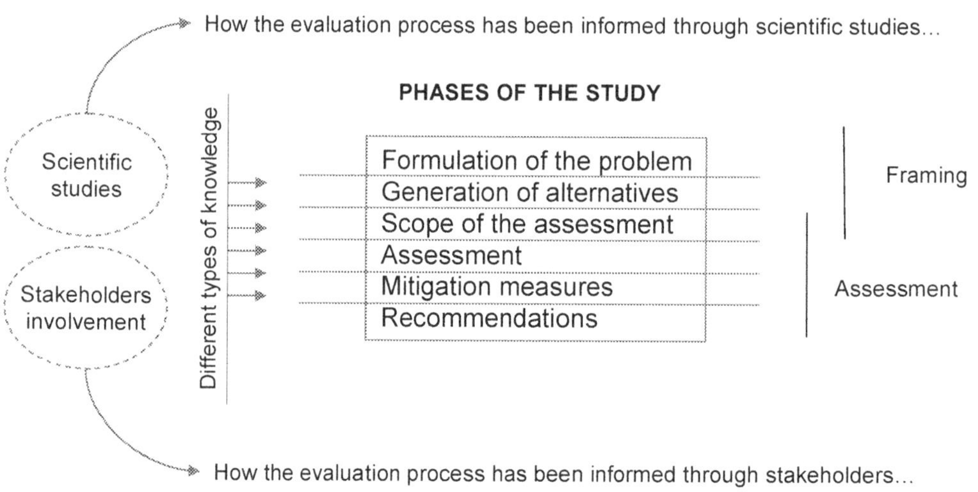

Figure 5.3 Different types of knowledge: scientific and non-scientific information. How and when were integrated in the assessment activities

Models used in environmental planning processes are not only evaluation tools but they should be considered as instruments to structure significant issues as well as to facilitate and to engage the relevant community in a dialog leading to the resolution of conflicts aroused in decision-making processes.

Quality assurance processes of models applied in problem-solving procedures should consider the following:

- the ability of the model to internalise the information, alternatives and criteria with minor loss of information,
- the recognition of the model and its results, by the scientific community, as well as by the agents involved in the process, and
- the transparency of the model, defined in terms of facility to handle and interpret results and the procedure's transparency itself.

Having these concepts in mind the following categories, **Fitness for Purpose**, **Legitimacy**, and **Transparency** will be set up to explore the quality of the decision-aid tools used in planning processes.

The model's Fitness for purpose will make explicit whether the methodology suits the issue under analysis. It tries to answer an important question (fundamental in the quality assessment of a tool or object), does the model fit the problem or is the problem being adapted to fit the model? This question is trying to be answered defining three categories: Adequacy, Accuracy and Information Encoding. The model's specificity will be analysed in terms of whether it was developed to deal with the specific type of issues or on the contrary it is a generic methodology used in several different spaces of decision-making processes. The adaptability of the methodology to the problem regards the way it may tackle non envisaged aspects. The Encoding Information refers to data processing issues, namely, the necessity to transform, to codify or to translate the information in order to be used during the model process.

Models not only have to match the issue under analysis, but they also should be 'certified' by the experts and accepted by the actors involved in the process. This acceptance by social actors is crucial in the case of decision issues where there are conflicts of interests and compromise solutions perhaps might be needed. The category Legitimacy addresses this idea through two aspects:

- the level of Acquiescence of such methodology by the Scientific Community, the stakeholders and the extended Community,
- the kind of Control Procedure (if it does exist) applied to the obtained results.

Decision-aid models (mainly those dealing with environmental matters) should be amongst other things, instruments to facilitate and to engage the whole Community in a dialog to promote the resolution of conflicts. The category Transparency is defined to assess the decision-aid model's ability to achieve this objective, exploring the understanding of the used methodology for both their internal mechanisms or procedures and the results generated through its application.

Background information (Scientific & Technical Information):

One of the requirements demanded to any kind of software, instrument or object is its **fitness for purpose** to reach the aims for which it was designed or applied. Similarly the information or the knowledge used in the planning process must fulfil that requisite. In QAAT, this adjustment is analysed on the basis of three attributes: Adequacy, Accuracy and Completeness. Adequacy shall reflect the sensitivity of the information to reach the envisaged aims, that is to say, their adequacy to the problem. Completeness is intended to examine whether the available data are complete or, on the contrary, they show some 'lacks' (i.e. incomplete series, stakeholder opinions not available...). At last Accuracy states the uncertainties related to the information analysed.

As important as the fitness for purpose is the **Applicability**; when the information cannot be used during decision-making processes, its quality will certainly be affected (independently of the degree of adjustment of the Information). Applicability is defined on the basis of Accessibility and of Understanding. Accessibility to the information and its sources plays an important role into the planning process -the information 'to be applicable has to be available'. Understanding reflects the intelligibility of the information by the different stakeholder, as well as by the community at large. When data are not comprehensible, situations where some variables or effects are less valued (or on the contrary valued in excess) may arise, affecting the quality of the decision process.

The Information used in problem-solving processes should be **Reliable**, mainly when dealing with environmental planning, where stakes are high, decisions are urgent (and in many cases irreversible) and there is irreducible uncertainty and ignorance. The assessment of Reliability will be implemented through the use of two complementary characteristics: Control and Confidence. The former relates to both the description of the sources and the process of verification of the information while the confidence deals with the degree of legitimacy of the data.

Those three attributes represent some of the quality characteristics that information used in river basin evaluation processes should fulfil.

Communication & reporting activities:

Communication and reporting activities quality assessment will entail the assessment of information used for or produced from these activities.

The criteria to evaluate information used for reporting and communication purposes will take into consideration not only information pedigree but also issues such as intelligibility (e.g. compliance with audience), style of reporting, management of uncertainties, etc.

5.4.3. Case Study: The Baixo Guadiana Participatory Management Process

This protocol (see Annex I) was applied in a normative and descriptive fashion within a participatory modelling activity for the Baixo Guadiana Management Plan. As a normative aspect categories were used to inspire the ways in which the mediated modelling process was conducted by the research team. The descriptive task consisted of proposing a questionnaire to the participants in which some of the quality categories of the protocol were included.

Other quality assurance categories dealing with more technical aspects were implemented by external experts. Preliminary conclusions derived from the application of this protocol are presented in Chapter 9.

REFERENCES

Commission of the European Communities (CEC) (2000) Directive 2000/60/EC of the European Parliament and of the Council establishing a framework for the Community action in the field of water policy, Published at Official Journal (OJ L 327) 22 December 2000.

Corral Quintana, S. (2000) "Una Metodología Integrada de Exploración de los Procesos de Elaboración de Politicas Públicas", Universidade de La Laguna, Tenerife, EC - Joint Research Centre: EUR 19724 ES.

Del Moral, L., Pedregal, B., Calvo, M. and Paneque, P. (2002) River Ebro Interbasin Water Transfer. Work Package 1 of project ADVISOR, Universidad de Sevilla.

Funtowicz, S. (2001) Peer Review and Quality Control. In International Encyclopedia of the Social & Behavioral Sciences, ed N. J. Smelser & P. B. Baltes, 11179-11183, Pergamon, Oxford.

Funtowicz, S, and Ravetz J.R. (1990) Uncertainty and Quality in Science for Policy, Kluwer, Dordrecht.

Gibbons, M. (1999), Science's New Social Contract with Society. *Nature* **402**: C81-C84.

Hatzilacou, D., Kallis, Giorgos and Coccossis, H. (2002) The River Evinos Reservoir, Work Package 1 of project ADVISOR, University of the Aegean.

Hill, G., Urama, K., Spash, C. and Wynn, G. (2002) The Designation of the River Ythan and Estuary as a Nitrate Vulnerable Zone, Work Package 1 of project ADVISOR, The Macaulay Institute.

Ravetz, J.R. (1996), Scientific Knowledge and its Social Problems, Transaction, New Brunswick NJ and London.

Van Leeuwen, E., Dalhuisen, J., Vreeker, R. and Nijkamp, P. (2002) The Grensmaas Project, Work Package 1 of project ADVISOR, Vrije Universiteit Amsterdam.

Videira, N., Antunes, P.; Borrego, D., Santos, R. (2004) Mediated Modelling for Integrated Water Resources Planning and Management – The Baixo Guadiana Experience, ADVISOR internal document, New University of Lisbon.

Videira, N., Lobo, G. Antunes, P., Santos, R. and Guimarães Pereira, Â. (2002) Alqueva Multipurpose Project, Work Package 1 of project ADVISOR, New University of Lisbon.

Chapter 6

Combined CBA and CEA Framework for Flood Protection

Frank Bruinsma and Ron Vreeker

Free University of Amsterdam, The Netherlands

6.1. INTRODUCTION

The main objective of ADVISOR is to provide an integrated project evaluation framework and methodology for the the sustainable governance of Europe's river basins, which will be achieved through the delivery and dissemination of a set of guidelines. This set of guidelines consists of a structured evaluation process, criteria and appropriate practical methodological tools.

As part of the activities in ADVISOR, a combined cost benefit analysis (CBA) and cost effectiveness analysis (CEA) framework has been developed. The framework is developed for the main Dutch river basins of the rivers Rhine and Meuse.

In recent years some serious floodings and threats of flooding occurred due to heavy rainfalls in amongst others France, Italy, Germany, Belgium and the Netherlands. This has led to an increased interest in improving the protection of, in particular, economic valuable areas. However, experiences with dike strengthening projects (by making dikes higher and wider) in the Netherlands in the 1970s had shown that such projects led to major interventions in the landscape and affected the cultural and natural values in the areas concerned. This resulted in an increased tension between safety and sustainability objectives; it seemed difficult to combine both objectives (safety and sustainability). The destruction of areas with high cultural and/or natural values by the extension of dikes led to major protests of local inhabitants and national environmental pressure groups. This was one of the first experiences the Dutch government had with public resistance based on environmental motives against policy measures aimed at improving safety.

Since then it is recognized by the Dutch government that in particular in flood protection projects, next to the major economic interests, one has to take environmental interests also into consideration. The economic impacts concern both the high costs of the protection measures to be taken, as well as the high economic benefits; the value of the properties protected against flooding. The environment impacts often concern the loss of high cultural and natural values in the concerned river basin areas.

Given the large economic interest it is obvious that in most evaluations cost benefit analysis is applied (see van Leeuwen, Vreeker & Bruinsma, 2003). Moverover, since the year 2000 large infrastructure projects in the Netherlands are by law obliged to be evaluated by means of cost benefit analysis (CPB/NEI, 2000).

As will be explained in section 3, the government's major concern is to assure certain safety levels. In a cost benefit approach the project with the highest NPV or cost-benefit ratio – given a certain level of safey – is deemed the most

 Integrated Evaluation for Sustainable River Basin Governance. Edited by Nuno Videira, Giorgos Kallis, Paula Antunes and Rui Santos. ISBN: 1843391481. Published by IWA Publishing, London, UK.

desirable. Section 3 shows that such outcomes are produced by projects which encompass only technical solutions, such as an increase in the height of a dike or the enlargement of a summer bed of a river. The technical solutions are in general the cheapest to achive the specified levels of safety. However, technical solutions are in general the least sustainable or environmental friendliest. Thus, in our evaluation framework, a two step approach is taken. In the first step a cost benefit analysis is applied based on the technical solutions each alternative offers. The second step concerns a cost effectiveness analysis in which, alternative partial, but more environmental sustainable solutions are tested, while keeping the level safety constant.

Cost benefit analysis and - to a lesser extent - cost effectiveness analysis have some clear shortcomings concerning the incorporation of environmental impacts and - in particular - stakeholder interests in the evaluation. Since environmental impacts are often brought to the fore by stakeholders, such as environmental pressure groups, there is a clear need to extend the both to cope with those interests.

The Waterframework Directive strives also for a better accommodation of stakeholder participation in planning and assessment procedures. In this chapter, we propose a combined use of CBA and CEA. In particular when partial, more sustainable, solutions are tested by CEA, emphasis is laid on the introduction of stakeholder interests in the applied research methodology.

In the next section we briefly discuss the advantages and disadvantages of CBA and CEA, in particular related to the evaluation of flood protection projects. In section 3 we give a description of the Dutch case study project 'Space for the River'. In section 4 we apply the cost benefit analysis for the Dutch project 'Space for the River', and in section 5 - based on the same case study - the cost effectiveness analysis will be explained. In the final section 6 we will propose an approach to add public participation in the evaluation process, based on the combined cost benefit analysis and cost effectiveness analysis framework. Finally, we will draw some conclusions in section 7.

6.2. DESCRIPTION OF CBA AND CEA METHODOLOGIES

CBA-methodology

The standard framework for evaluating policies and investments from an economic perspective is Cost-Benefit analysis (CBA). The French engineer Dupuit has laid the theoretical foundation for Cost-Benefit analysis in 1844. Cost-Benefit analysis is an approach to policy recommendation that permits analysts to compare and advocate policies by quantifying their total monetary costs and monetary benefits (Dunn, 2003). The social cost benefit analysis model is based on the theory of welfare economics, according to which the welfare of a society depends on the aggregate individual utility levels of all members of that society (Brucker et al. 1998). The aim of cost-benefit analysis is to locate and include all effects of a project and to express the unobservable change in welfare in observable monetary units. A project increases the welfare of a society if the winners can - hypothetically - compensate the losers, while still attaining a higher utility level than was the case prior to the project (Hicks-Kaldor compensation or potential Pareto improvement).

CBA typically uses consumer surplus as the operational measure for benefits. The advantage of consumer surplus is that it is based on directly observable market demand functions by taking individuals' willingness to pay as the basis for the determination of benefits.

In carrying out a CBA various steps are taken. In the first step, the project and its alternatives are defined, along with the relevant population (those individuals whose welfare is considered relevant and affected). In the second step, the relevant impacts of the project and its alternatives are identified, using the criterion of whether they constitute a cost, a benefit or a redistribution of welfare. Furthermore, all stakeholders should be identified that will be positively - beneficiaries - and negatively - target groups - affected by the adoption and implementation of the policy recommendation. In the third step, relevant costs and benefits are valued in monetary units. Where price changes are involved, this necessitates calculation of changes in consumers' and producers' surplus. For changes in the supply of public goods estimates of compensating or equivalent variations are necessary. When no market prices exist to value costs and benefits, non-market valuation techniques are used. Costs are measured as social opportunity costs. Where market imperfections or market interventions create differences between market prices and marginal social costs or benefits, shadow prices are calculated. In the fourth step, benefits and costs are aggregated over time by means of discounting. Discounting expresses future costs and benefits in terms of their present value. The correct discount rate to use in CBA is a matter of great dispute, as capital market imperfections drive a gap between the social time- and the social opportunity cost of capital. In the Netherlands, the Ministry of Finance (1995) prescribe a discount rate of 4% for projects of national importance. The fifth step of a CBA involves the comparison of total discounted benefits with total discounted costs, to produce a net present value (NPV). If the net present value is positive, then the project has passed the CBA test. Alternatively, information may be presented in terms of benefit-cost rations; or as an internal rate of return, which is the discount rate that yields a zero NPV. The final step in a CBA consists of conducting a sensitivity analysis on important parameters such as the discount rate, project life-span, and cost and benefit estimates.

However, the application of CBA encounters several problems which can be summarized as: 1) distorted markets or 'second best' conditions; 2) valuation of environmental effects; 3) determining the discount rate; 4) spatial scope of CBA; 5) risk and uncertainty; and; 6) equity concerns. Considering the evaluation of flood protection projects the problems 2, 4, 5, and 6 are of particular interest. Those problems will be discussed more in detail below.

Valuation of environmental effects
The inclusion of environmental effects in a CBA requires the economic (monetary) valuation of environmental goods. Since no markets for environmental goods exist, it is difficult to observe market prices that reflect marginal costs or benefits.

Therefore, various valuation methods have been developed that aim to identify shadow prices for these goods. Various types of environmental values that can be distinguished. Nijkamp et al. (2002) provides us with an overview (see table 6.1).

Table 6.1 A classification of environmental values (Source: Nijkamp et al. 2002)

<table>
<tr><td colspan="4">Use Value
(related to the 'use' of the good as a consumption good or production factor)[13]</td><td colspan="4">Non-use value
(not related to the 'use' of the good as a consumption good or production factor)</td></tr>
<tr><td colspan="2">Direct use value
(the use of the good itself)</td><td colspan="2">Indirect use value
(the existence of the good improves the quantity or quality of other consumption goods or production factors)</td><td rowspan="2">Option value
(related to possible future use of the good)</td><td rowspan="2">Quasi-option value
(related to possible future information on the importance of the good)</td><td colspan="2">Existence value</td></tr>
<tr><td>As a consump-tion good</td><td>As a production factor</td><td>Affecting consumption goods</td><td>Affecting production factors</td><td>Philan-thropic value (for contem-poraries)</td><td>Bequest value (for future genera-tions)</td></tr>
</table>

Valuation methods aim to estimate the individuals' marginal 'willingness-to-pay' (WTP) in monetary terms for an improvement in the quantity or quality of the environmental good concerned, and therefore the methods are consistent with the general philosophy of CBA. A brief outline of possibilities to derive estimates of the value of environmental effects is set out here. Table 6.2 gives a concise overview of the valuation methods used for environmental valuation (see for example, Munasinghe, 1992 and Verhoef, 1996).

In table 2.2, market or behavioral based valuation methods are depicted on the left-hand side. Two main categories of behavioral techniques can be distinguished. Revealed preference techniques can be applied when surrogate markets for the environmental good to be valued exist; that is, when consumers' marginal willingness to pay for changes in the effect can be measured by looking at their behavior on other, related markets by means of hedonic techniques, travel cost approaches or household production functions.

When the goal is to value non-use values, or when no surrogate markets exist, stated preference techniques could be used to infer consumers' willingness to pay by confronting them with hypothetical markets. Contingent valuation studies try to ask for a willingness to pay directly, possibly by confronting respondents with two (or more) scenarios for which the quantity or quality of an environmental good and some financial transfer vary, and asks them to indicate the most preferred option.

Table 6.2 An overview of classification of valuation methods (Verhoef, 1996)

<table>
<tr><td colspan="3">Valuation approaches</td><td colspan="2">Short-cut approaches</td></tr>
<tr><td colspan="2">Behavioral</td><td colspan="3" rowspan="2">Non-behavioral</td></tr>
<tr><td>Surrogate markets (revealed preference)</td><td>Hypothetical markets (stated preference)</td></tr>
<tr><td>Hedonic techniques
Travel cost methods
Household production functions</td><td>Contingent valuation
Conjoint analysis</td><td>Damage costs
Costs of illness</td><td>Prevention costs: hypothetical defensive, abatement or repair programs</td><td>Actual defensive abatement or repair programs</td></tr>
</table>

[13] Note that 'consumption' needs not always deplete the good: enjoying a scenic view is also a form of consumption.

These valuation techniques show that despite the lack of a market for environmental products and services, tools exist to value environmental effects.

However, some limitations do exist. A first limitation concerns *absolute imperfect information* (Van Pelt, 1993). In order apply a valuation method to come to a proper valuation of effects, environmental effects need be assessed in physical terms. Due to the complexity of environmental problems, however, a significant degree of uncertainty is likely to remain in many studies.

The WTP principle underlying all valuation techniques constitutes a second limitation. *Relative imperfect information* hampers comprehensive valuation by individuals. Due to a lack of awareness, people may appear to be willing to pay relatively little for the environment (Hueting, 1980). Furthermore, no method is available to account for the interest of future generations that may be affected. This problem particularly refers to option, bequest and existence values and to a lesser extent to direct and indirect use values.

A third limitation is that most valuation techniques are *partial approaches*, focusing on a single aspect of environmental quality. Whereas the travel cost method is especially useful to value recreation areas, the cost of earnings technique is applicable if human health is affected. Because most techniques focus on environmental effects that can be linked to existing markets, they may fail to incorporate other values resulting in an inadequate valuation of environmental effects.

A fourth limitation is *method uncertainty*. Apart from method-specific shortcomings, several studies have shown that if a number of valuation techniques are applied in a certain case, (significantly) different results may be obtained. A related, serious problem in the case of CVM (contingent valuation method) is the divergence between WTP (willingness to pay) and WTA (willingness to accept) measures of value.

Finally, the results of methods may raise criticism from an *equity* point of view (Van Pelt, 1993). As CBA is indifferent with regard to existing income distributions, so are WTP-indicators for the environment. The concept of WTP may also be inappropriate in the light of intertemporal equity.

When resources are lacking to undertake a 'market based' valuation study, short-cut approaches are sometimes used. These techniques are often criticized on many grounds. Important considerations are that circularity may plague the estimated values (e.g., an externality is considered unimportant by the authority would induce small public defensive outlays, which would subsequently produce a low estimate of the external costs), that defensive outlays may not always be possible, that defensive outlays by private parties are typically carried out up to the point where marginal benefits equal marginal costs so that external costs remain in existence that are not measured using the method, and that hypothetical 'full defense' programs may on the one hand consider too large a reduction of the externality, but on the other hand bear no relation to the true external costs if they are not actually carried out.

Spatial scope of CBA

Projects may affect the welfare of people living nearby, but also of people further away. From the welfare economic literature it is well-known that the impacts on economic welfare may include substantial *distributive* effects in addition to the *generative* effects. As a result, a project's desirability, as well as the ranking of alternatives, may depend on the definition of the 'planning area' and, with that, of the (spatial) set of individuals whose welfare is assumed to matter in the CBA.

The geographical boundaries for a CBA may in some instance be defined by the jurisdiction that commissions the appraisal, and that would finance the prospective project(s). In other cases, official guidelines may exist that for instance stipulate the appropriate level of spatial aggregation should be the national one, for instance motivated by the desire to avoid wasteful investments by level authorities that would merely redistribute – rather than generate – national welfare. From an overall efficiency viewpoint, it is rather obvious that everybody would eventually benefit if all CBA's would take a global perspective and hence consider welfare impacts 'world-wide', and – lump-sum – redistributions between and inside jurisdictions were possible (Nijkamp et al. 2002). But it is probably a rather theoretical consideration that may not be viable in the practice of decision-making. In any case, the spatial demarcation is an important aspect for the design and interpretation of CBA's.

Risk and Uncertainty

CBA has traditionally had difficulties with coping with risk and uncertainty. However, tools that are offered to account for risk and uncertainty may have significant limitations in the case of environmental effects. The usual recommendation is to apply sensitivity analysis. This approach, showing how close a project is to marginal feasibility levels, is not very useful in accounting for gaps in scientific knowledge about ecosystems. Another drawback is that sensitivity analysis usually is a partial approach, which fails to address (uncertainty about) linkages between ecological variables. Finally, rarely is sensitivity analysis linked to ecological risk attitudes of decision-makers.

A second mechanism to account for uncertainty involves the calculation of expected values for ecological and other variables. However, expected value approaches only address the problem of risk (i.e. all possible events are known) and require insight into probability distributions for environmental events, which may frequently be lacking.

Information required to calculate expected costs will, in general, be harder to obtain if non-linearities and discontinuities characterize the behavior of the ecological system (Hanley, 1999). In practice, CBA's may partially account for this by deploying scenarios and sensitivity analyses, and environmental policy may be based on precautionary principles that on the one hand may prevent huge irreversible future environmental cost, but on the other hand may lead to considerable current costs or foregone benefits. Regrettably, under this type of uncertainty, by definition no 'optimal degree of precaution' can be established. It is evident that the existence of uncertainty may substantially reduce the transparency of CBA's.

Equity concerns

In the previous section attention was devoted to the inclusion of environmental effects in CBA. However, distributions of impacts should also be accounted for in the analysis.

In principle there are two ways to account for equity objectives in evaluation processes. The first is social CBA, whereby through quantitative weights outcomes of economic CBA are adjusted for income distribution objectives.

The second possibility is to score project alternatives on a separate equity criterion. Sometimes it may be possible to develop quantitative indicators, which usually are based on the same philosophy underlying (in)equality measures for countries, such as the Lorenz curve and the Gini concentration ratio.

At the project level less data-demanding indicators are needed. Focusing on relative income, an indicator may be taken as the ratio of the part of net efficiency benefits that accrue to target groups and the total net efficiency benefits. Alternatively, changes in absolute poverty measures may be used as scores on the equity criterion. This requires the definition of a poverty line, representing minimally acceptable levels of income or certain groups of basic needs. To overcome the data requirements of the above-mentioned indicator, some new tools for equity analysis are developed. Lichfield, for example, has developed the Planning balance sheet method (Lichfield, 1964) to integrate distributive impacts in assessment. This approach involves the construction of a matrix with the various alternatives on the one side and possibly affected population groups on the other side.

CEA - methodology

The cost effectiveness analysis is closely related to cost benefit analysis. Both methods try to quantify the pros and cons of the proposed project alternatives. In contrast to cost benefit analysis, which attempts to measure all relevant factors in a common measurement unit (monetary values), cost effectiveness analysis uses two different measurement units. Costs are measured in monetary units, while effectiveness is typically measured in units of goods, services, or some other valued effect (Dunn, 2003). In the absence of a common unit of value, cost effectiveness analysis does not permit net effectiveness or net benefit measures, since it makes no sense to subtract the total costs the from total units of goods or services. It is however possible, to produce cost effectiveness and effectiveness cost ratios, for example, ratios of costs to units of preserved nature areas. These ratios have a different interpretation than cost benefit ratios. Whereas effectiveness cost and cost effectiveness ratios tell us how much of the good or service produced per Euro expended - or alternatively, how many Euro are expended per unit produced - benefit cost ratios tell us how many times more benefits than cost are produced in an given instance.

Cost effectiveness analysis is particularly appropriate for questions involving the most efficient way to use resources to attain objectives that can not be expressed in terms of income. Cost effectiveness analysis has been used to recommend alternative policies and programmes in for instance criminal justice, transportation and health. Cost effectiveness analysis is well suited to the analysis of externalities and intangibles, since these types of facts are difficult to express in Euro as a common unit of measure. The strengths of cost effectiveness analysis are in its comparative ease of application, its capacity to treat collective and quasi-collective goods whose values cannot be estimated on the basis of market prices, and its appropriateness for analysing externalities and intangibles. The main limitation of cost effectiveness analysis is that recommendations are less easily related to questions of aggregate social welfare (Dunn, 2003).

6.3. PROJECT DESCRIPTION OF THE CASE STUDY: SPACE FOR THE RIVER

The project Space for the River concerns the development of a new spatial policy of the national government for the development of the main river basin in the Netherlands (Ministry of Transport and Water Works et al. 2002). In this project there are two central themes: 1) protection of the river areas against floods and 2) measures to improve the sustainability situation (preservation of landscape, nature and cultural inheritance) of these river areas.

In the project a number of alternative development scenarios will be propound. Those scenarios will be tested in a combined cost-benefit analysis and cost effectiveness analysis framework, in which there is, however, no particular emphasis for the interests of stakeholders. Nevertheless, the adding of a cost effectiveness analysis to the evaluation framework opens the door to a successful implementation of some forms of stakeholder involvement, in particular at the local level (see section 6).

There are two major river systems in the Netherlands involved in the project Space for the River; first and most important the system of the river Rhine and second, the system of the river Meuse (see figure 6.1).

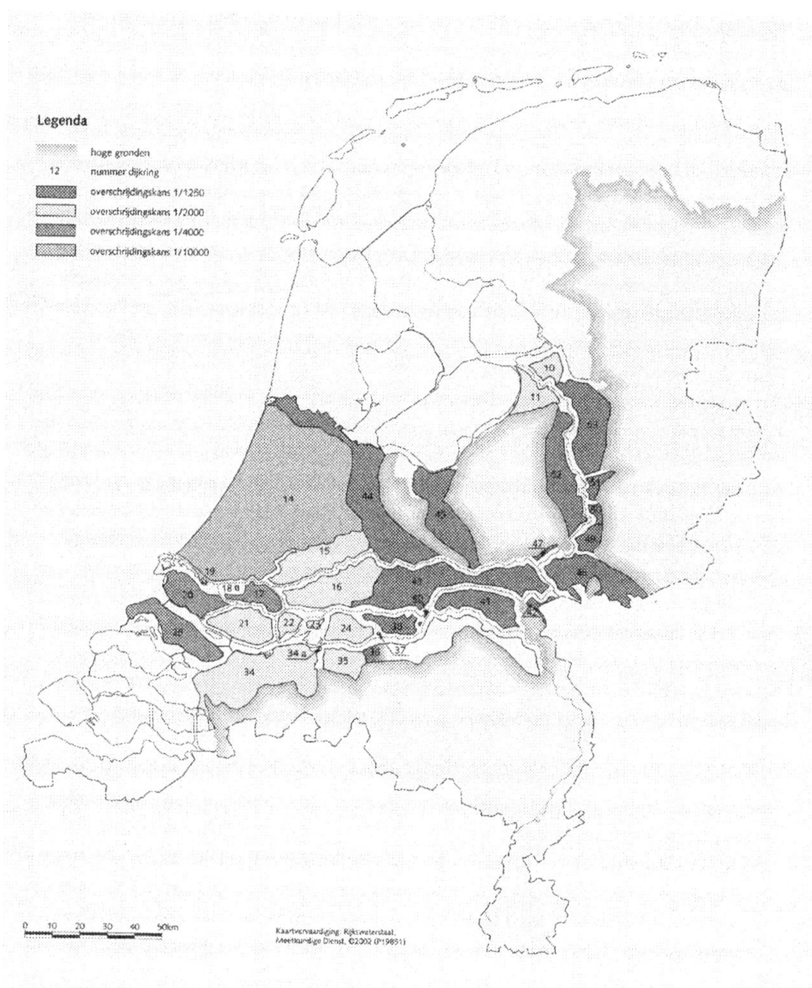

Figure 6.1 The Dutch river systems and the risk of flood (Source: Ministry of Transport and Water Works et al. 2002)

A third large river is the river Schelde, but this only flows through a small part of the country. Nevertheless, the last river is important because it determines the accessibility of the Belgian harbour of Antwerp.

The water of the river Rhine flows via the IJssel into the lake IJsselmeer (the former sea Zuiderzee) and via the Lek and Nieuwe Waterweg or via the Waal, Merwedes and Hollands Diep into the North Sea. The water of the river Meuse flows directly via the Hollands Diep into the North Sea.

After the floodings of 1993 and 1995 it has become clear that the protection against flooding has to be improved. In order to cope with high water outlets running through, the outlet capacity of the river Rhine and Meuse has to be increased from 15,000 m^3/s to 16,000 m^3/s and 3.650 m^3/s to 3.800 m^3/s, respectively. In 2030 this figures are expected to have risen to 18,000 m^3/s and 4,600 m^3/s, respectively.

A second threat in the deltas of those rivers is the rise in the water level of the North Sea in general and the rise of the sea water level during storms in particular. The overall rise in the sea water level is expected to be 60 cm in the long term.

In the past, next to the construction of dikes, most measures applied tried to improve the speed of the water flow to increase the water outlet capacity of the rivers (such as short cuts of meanders). Furthermore, measures were taken to improve the ability of rivers for inland shipping. However, urban development took place near rivers or even within the winter bedding of the river. Due to these urban - economic - developments the rivers got less space to carry off water, and the risks and size of the impacts of flooding increased over time.

The second target of the Space for the River policy, to improve the sustainability situation of the river areas, is caused by the increasing spatial claims by functions such as housing, infrastructure and industrial activities in those areas. These spatial claims constitute a major threat for the open character of the landscape, such as; urbanisation, infrastructure, and industrial sites. Sustainability has to been seen from the social (cultural diversity, spatial diversity, attractiveness, social justice), economic, and ecological perspective.

Since safety and sustainability targets are hard to combine, there exists a tension between them. As already mentioned in section 1, the dike strengthening projects in the 1970s have led to severe damage to the cultural and natural landscapes in river basin areas in the Netherlands. The main target of the Space for the River project concerns safety. Although, sustainability is an additional objective, this does not mean that safety is the only target that counts. The quantitative safety targets have to be achieved; however, both targets - safety and sustainability - stipulate where space is available and which measures have to be taken.

The main strategies to improve the safety and sustainability proposed in the project Space for the River are a mix of technical and spatial planning measures. The following strategies are proposed:

- Retain water in such a way that it flows less fast into the river;
- Reduce peaks in water flows by temporally storage;
- Offer more space to the rivers.

It is important to note that the choice of which strategy could be applied best, and the effects of those selected strategies, in the Netherlands strongly depends on the strategies applied in the countries up stream the river. If the inflow of water in the Netherlands is more constant (due to a reduction in the peak flows) in the future caused by measures taken in, for instance, Germany, the need for strategies aiming to reduce peaks in water flows by temporally storage become less important.

As stated above those strategies can be achieved by numerous combinations of technical and spatial measures. Those technical and spatial measures are presented in Table 6.3.

Given these strategies, and the technological and spatial measures, various scenarios can be developed. Those scenarios are based upon given a number of priorities, from the perspective of spatial quality and sustainability. First, the solutions have to be found within the river bed itself (the first four spatial measures mentioned in table 6.3). Only when measures within the river bed are insufficient to reach the targets of safety and spatial quality, or when the costs of those measures are too high, the second or third priorities come to the fore. The second priority concerns large scale measures outside the river bed (the fifth till seven spatial measure mentioned in table 3.1). Those measures have a large impact on the inhabitants of those areas and can only be realised over a long time period. The third priority concerns the technical measures mentioned in table 6.3. Those measures are difficult to combine with the sustainability targets of the policy Space for the River. The lowest priority is given to the diversion of river water to other river systems.

Table 6.3 Possible spatial and technical measures

Spatial measures	Deepening of the forelands of the river Construction of parallel watercourses Removal of obstacles, such as bridges, roads, ferry steps, etc, from the winter bed of the river Enlargement of the forelands by replacement of dikes Construction of retention areas for temporally storage Construction of temporally by-passes and green rivers (alternative rivers only used in periods of high water levels) Reduction of the inflow of water from small local or regional rivers Divert water to other river systems
Technical measures	Lowering of groynes in the river Enlargement of the summer bed of the river Increase in the height of the dikes

To conclude, from the perspective of spatial quality and sustainability, the technical measures have the lowest priority. However, from the cost perspective the technical measures often have the highest priority. In most cases it is by far cheapest to apply technical measures to arrive at the desired levels of safety.

In the project Space for the River a number of scenarios are developed that sketch the measures to be taken to arrive at the specified safety levels and taking spatial quality and sustainability aspects into account to a maximum. The scenarios developed are subject to a number of preconditions:

- the prevailing safety standards;
- the present division of the water flow (65% of the Rhine water via the river Waal, 20% via the river Lek, and 15% via the river IJssel);
- the hydraulic preconditions;
- the inflow of local and regional rivers remains unchanged;
- the existing legal framework;
- the function of the rivers for inland shipping remains unchanged.

Next to those binding preconditions, there are a whole series of non-binding basic principles. The main basic principles are:

- optimisation of the societal support for the scenarios;
- a long-term vision for the future of the river basin area;
- the existing policies concerning spatial planning, nature, recreation, landscape, environment, and water;
- international agreements considering the prevention against high water levels;
- the opportunities of extracting sand and gravel;
- existing water projects;
- etc.

As shown in the list above, the optimisation of the societal support is one of the main non-binding principles. To assure the societal support, stakeholder participation is an important instrument. Finally, the scenarios will be judged on the following impacts:

- protection against flooding;
- river morphology (the shape of the river and the river bed). The river morphology defines the speed of the water flow and the sedimentation of materials. Thus the river morphology has major impacts on the maintenance (for instance dredging and the stability of dikes and other construction works) and the life span of the waterworks;
- soil and water. The creation of additional space for the river may lead to the extraction of polluted soil or opposite sand and clay that can be used for industrial purposes. The financial impacts must be calculated. Also the impacts of the proposed projects on the quantity and quality of the water might be considerable. It might lead to a further penetration of salt sea water in the river deltas, and the construction of for instance bypasses may lead to major interventions in local water management systems. Furthermore, the projects may lead to changes the groundwater characteristics (level, quality, quantity, and direction);
- nature impacts;
- landscape impacts;
- cultural historical values and archaeology;
- inland shipping (the river Waal is the most intensive used river for inland shipping in Europe);
- agriculture (the main land use function in the Dutch river basin areas);
- recreation;
- social economic impacts (economic sectors such as industry, agriculture, and recreation);
- the future standing;
- mitigating and compensating measures.

This project Space for the River will be used to clarify the combined cost-benefit analysis and cost effectiveness analysis framework we propose to evaluate policy recommendations considering safety and sustainability aspects in large river basin projects.

6.4. COST BENEFIT ANALYSIS

The purpose of social cost benefit analysis is, from a welfare economic perspective, to compare a number of alternative scenarios on the basis of monetarised costs and benefits. In general, government actions should add to the welfare of the citizens. Public expenditure in project A is justified only after it has been shown that project B, C, D, etc, will lead to a smaller improvement of public welfare. This rule also holds for public expenditures in protection measures to reduce the chance of flooding. The costs of protection measures should not exceed the benefits (for instance, avoided damage costs - economic as well as nature, landscape, and cultural historic damage - and victim's costs) of the project. In a number of studies in the Netherlands it is shown that technical measures are the least expensive to fulfil safety targets.

Since safety is the main target of the Space for the River project, the analysis will be developed in two steps:

- In the first step - based on cost benefit analysis - the benefits of higher safety levels are compared with the costs of the measures taken to attain those safety levels. Those safety levels will differ for the various dike rings, depending on the economic value within the dike ring. For instance, given the economic value of the Randstad area, the chance of a flood in this area of once in every 10.000 years is acceptable, whereas in more agricultural areas the chance of once in every 1.250 years is acceptable. Even more, in the Dutch policy it is discussed that some upstream areas must be prepared to be flooded on purpose (the so-called retention areas) to protect downstream areas with high economic values. As stated above, it is expected that technical measures are the most cost-effective way to fulfil the objectives. In the first step the costs of technical measures are compared with the benefits of the project.
- In the second step - based on cost effectiveness analysis - alternative scenarios are developed by adding packages of measures to improve the sustainability situation of the area, given the fixed safety levels. In those scenarios the additional costs to improve the spatial quality of the area are compared to the benefits of the improved spatial quality. Additional benefits could occur when sand, clay, and gravel for industrial purposes are extracted. However, the problem is to define useful packages of measures to improve the sustainability of the area. In principle, an endless list of combinations for different parts of the area is possible. Therefore, it is important to analyse the simple cost benefit analysis considering only technical safety measures first as a benchmark.

The initial Cost-Benefit analysis framework is based on technical solutions, in particular, the increasing of the height of the dikes. This starting point for CBA is chosen from a cost perspective: the costs of increasing the height of the dikes to arrive at the specified safety levels are relatively low compared with alternative solutions. This seems to be a simple, straightforward approach. However, there are a number of factors that make this cost benefit analysis rather complex:

- uncertainty about the height of the damage resulting from flooding. One has to make division between the damage costs that can be monetarised and damage costs which are difficult or impossible to monetarise. The first category contains direct damage to buildings, infrastructure, cars, stock, etc, and indirect costs such as the damage caused by interruption of the production process, or the costs of the impact of the flooding in area A on the production process in area B. Finally, by improving the safety against flooding the behaviour of people will change. The reduced risk will lead to additional investments and thus overtime to an increase in the economic value within the area protected against flooding. The second category - more difficult or impossible to monetarise costs - contains damage to nature, landscape, and human victims;
- the exact relation between the height of the dikes and the chance of flooding is unknown. Next to the height of the dike, also the strength of the dike and the stability of the underground are important factors influencing the chance of flooding. Recently, a few dikes built on peat collapsed during a period of drought in the Netherlands. The peat dried out and became unstable, due to natural gasses in the peat. The dikes broke and slide aside. This led to the flooding of residential areas in a period of extreme drought;
- the inflow of water will change overtime. Moreover, the sea level will rise, whereas the surface in the western part of the country sinks. Furthermore, it is not known which measures will be taken in the countries upstream the river, and by that it is not known by which speed and by which quantities the inflow of water will change;
- the project contains a number of mutual related solutions and measures. As mentioned before, due to the existence of the dike rings, if upstream an area is flooded, this has a positive impact on the chance that a downstream area will survive. This might lead to strategic choices in which agricultural areas upstream (with low economic values) are sacrificed for urban areas (with high economic values) downstream;
- the safety targets can be achieved by a number of technical measures with different impacts on the surrounding (and thus the sustainability situation of the area).

The uncertainties mentioned above make it difficult to define a straightforward safety scenario. Therefore, a rather pragmatic approach to define the safety scenario is proposed in which an increase in the height of the dikes is the standard solution from a cost-safety perspective. However, when other solutions are evident, these solutions will be applied. For instance, a green rive bypass, a bypass which is dry in periods with normal water outlets, might be preferred above the increase of the height of the dikes in cities.

For the safety cost benefit analysis the data requirements are the following:

- monetary and non-monetary values of the protected aspects within the dike rings;
- economic and demographic developments within the dike rings for the coming century;
- estimated damage of flooding;
- prognoses regarding river water inflow, increase sea water level, and the sinking of the surface in the next century;
- the chances of flooding per dike ring without the proposed measures;
- formulation of packages of measures to reduce the chance of flooding;
- calculation of the costs of the measures;
- impact on the chance of flooding of the measures.

Monetary and non-monetary values of the protected aspects within the dike rings

There are four types of damage:

- direct monetary damage to capital goods, for instance damage to real estate, capital goods, furniture, and damage due to loss of stock (products, raw materials and harvest);
- direct monetary damage due to the interruption of the production process (the decrease in value added due to non-production);
- monetary damage outside the dike ring (the interruption of the production process outside the dike ring, and the travel time losses due to detours since routes through the dike ring areas are cut off);
- non-monetary damage (victims, evacuees, cultural historical values, nature, environment, etc).

Economic and demographic developments within the dike rings for the coming century

Since the impact of the safety measures has to be calculated far into the future, information on economic growth, sectoral development, population growth, etc are required to estimate the potential damage costs for the future.

Estimated damage of flooding

With the above described information it is possible to estimate the damage for all the individual dike rings and for each year in future. Considering the costs of evacuation, the problem arises already before the situation becomes critical. So that evacuation costs might already be considerable without an actual flooding. In 1995 in the Netherlands 70.000 people within one dike ring were evacuated as a precautionary measure. In that particular case the dike withstand the water pressure. Nevertheless, the costs of evacuation were considerable.

Prognoses regarding river water inflow, increase sea water are level, and the sinking of the surface in the next century.

There still exists uncertainty about these issues. To cope with such uncertainties sensitivity analysis need to be applied.

Chances of flooding per dike ring without the proposed measures

To be able to arrive to an estimation of the damage costs one has to know the difference in the chance of flooding with and without the measures taken.

Formulation of packages of measures to reduce the chance of flooding

The cheapest way to attain the safety targets is to increase the height of the dikes. At present the flow capacity is calculated at 16.000 cubic metres per second for the river Rhine and 3.800 cubic metres per second for the river Meuse. However, it is expected that those capacities will rise to 18.000 cubic metres per second and 4.600 cubic metres respectively, in the far future. It has to be calculated if it is cheaper to immediately improve the dike rings to those capacities required in the future, or that a two-step approach (step 1 increase height of the dikes to the actual capacity requirements and - let us say - in about 50 years step 2 to arrive at the safety levels of 18.000 and 4.600 cubic metres) is cheaper. Furthermore, strategic choices have to be applied. Will areas with a low economic value be less protected then areas with a high economic value?

Calculation of the costs of the measures

Considering the calculation of the costs of the measures it should be clear what the public spending is, and which societal costs and business economical costs are involved.

Impact on the chance of flooding of the measures

Per dike ring the reduction in flooding chance has to be calculated.

When all information is collected a cost-benefit framework can be constructed, containing the elements presented in table 6.4. As can been seen most aspects can be measured in monetary terms. However, in particular aspects considering spatial quality are specified in qualitative terms. By replacing sub-projects based on technical measures by more sustainable sub-projects based on spatial measures to improve the spatial quality without changing the safety levels, the incomparability between alternatives in the project will increase (also the number of alternatives will rise exponentially by adding spatial measures as sustainable solutions). Therefore, a more tailor made approach has to be applied, that is better able to cope with qualitative information. In the second step of our proposal for an evaluation framework for river basin projects we add cost effectiveness analysis to the cost benefit analysis.

Table 6.4 Framework for cost-benefit analysis

Costs	Benefits
Costs of the measures (investment, maintenance, process costs and costs of infrastructure on dikes)	Monetary and non-monetary values of the protected aspects. Prevention of: direct monetary damage to capital goods, for instance damage to real estate, capital goods, furniture, and damage due to loss of stock; direct monetary damage due to the interruption of the production process; monetary damage outside the dike ring; non-monetary damage (victims, evacuees, cultural historical values, nature, environment, etc).
Monetary and non-monetary value of: destructed or relocated houses and industrial buildings, loss of agricultural land and recreational areas, and loss of living / working environment; production and hindrance of traffic in the period of construction; damage to nature, cultural history, landscape, ground water system, spatial diversity, etc.	

6.5. SOCIAL COST EFFECTIVENESS ANALYSIS

Description of the problem

After the cost benefit analysis based on safety targets, alternative scenarios have to be developed in which there is more emphasis on sustainability aspects, given the fixed levels of safety. Alternative packages of measures will be formulated of which the increase in spatial quality is compared to the additional costs. As stated before, next to the increase in spatial quality, there might also be benefits due to the extraction of sand, clay, or a gravel for industrial purposes. These benefits could come to the fore when spatial measures such as deepening of the forelands of the river, and construction of parallel water courses, are included.

The basic principle of cost benefit analysis is that all information is presented in one unit: money. However, considering sustainability it is difficult to express all impacts in this unit. What are the monetary benefits of the construction of a natural resource, or what are the costs of psychological damage in case of evacuation? Nevertheless, all impacts should at least as far as possible be quantified (in any possible unit). Of those impacts that only can be expressed in qualitative terms, it should at least be stated that this impact has a positive or negative sign. The final decision which the alternative scenario should be applied, remains a political choice.

The least cost benefit analysis can show is which price has to be paid for nature development, etc, in the alternative scenarios. However, since a number of impacts are not quantified in monetary terms the cost benefit analysis for the alternative scenarios has the character of a cost effectiveness analysis.

Which additional impacts can be distinguished?

- Costs. There will be investment costs and maintenance costs. Furthermore, there might be compensation costs, for instance for people and firms who are obliged to relocate in case of the creation of nature reserves. Finally, there might be costs caused by the dredging of polluted soil from the river bed. However, there might also be additional financial benefits by the extraction of sand, and clay, and gravel for industrial purposes.
- Inland shipping. Measures can have impact on the depth of the river, which could lead to additional dredging or a decrease in safety for inland shipping.
- Drinking water, groundwater. Measures can have impact on the availability of drinking water and on the level of groundwater. The level of groundwater may have impacts on agriculture and nature.
- Agriculture. Land may be withdrawn from agricultural use or there will be a partial loss of agricultural productivity. This might have impacts on supplying and purchasing industries.
- Business. The number of firms may have to relocate or end their activities.
- Recreation. Some recreational facilities may have to relocate or end their activities. On the other hand, new recreational opportunities might be created.
- Inhabitants. Some houses might need to be left and destructed. This might lead to capital costs (capital destruction) and social costs (due to involuntary relocations). Furthermore, the living environment might

increase (due to the creation of a nature reserve) or a decrease (due to the building of the dike in front of the house).

- Urbanisation. Most spatial measures will make further extension of the urban area impossible. In some occasions such an extension would have been a realistic option before the proposed measures.
- Infrastructure and other capital goods. Due to the measures infrastructure will have to be relocated.
- Raw materials. As stated before, extraction of raw materials may lead to additional benefits however, in some circumstances those raw materials and can be stored in depots first
- Nature.
- Cultural history.
- Landscape.

A number of those impacts have to be analysed in a broader context. Nature, cultural history and landscape also have to be analysed together to arrive at additional insights on their mutual relationships, diversity, and coherence with their surrounding.

In table 6.5 the additional effects of the cost effectiveness analysis as described above are given and translated in units of measurement. As shown in the table in particular the effects describing the spatial quality and sustainability, such as nature, cultural history, landscape, and spatial diversity are measured in qualitative units. Therefore, a cost effectiveness analysis is preferred above the cost benefit analysis to evaluate the impacts of spatial measures in the project Space for the River.

Framework for public participation in cost-effectiveness analysis

How can spatial quality as the equivalent of sustainability be defined and operationalised? Spatial quality is an interdisciplinary term, however, defined different in those disciplines. The concept of spatial quality has at least a spatial, a temporal, and a social dimension. Considering the spatial dimension both site (aspects of the location itself) and situation (the interaction with other locations) are important. Considering the temporal context, spatial quality refers to the impact on the spatial structuring of the present situation but also on the spatial quality for the far future situation. Finally, the spatial quality must fit within the societal context of political processes, market processes and the interests of individuals (Hooimeijer et al. 2001). Spatial quality is often defined by three values: user value, experience value, and future value.

Spatial quality is no well-defined economic term. Instead of spatial quality economists are used to terms like social welfare to express spatial quality. Dimensions of spatial quality are amongst others; nature, landscape, living environment, recreation, diversity, and spatial coherence.

We propose a rather pragmatic two step approach to operationalise spatial quality in river basin project evaluation. This two step approach is based upon the Space for the River policy (see also Renes et al. 2003). In the first step spatial quality is analysed at the local level. In the second step spatial quality is analysed at the interlocal, regional, level to check for spatial diversity and spatial coherence.

In the first step, the desired spatial developments in fields of nature, landscape, living environment, and recreation have to be inventoried at the local level. In the second step those locally desired spatial developments have to be evaluated at the regional level to test on spatial diversity and spatial coherence. This regional evaluation might lead, on the one hand, to additional synergy effects due to links between the desired local spatial developments. However, on the other hand, it might lead to the rejection of some of the local desired developments because those developments do not fit in the regional context.

Table 6.5 The additions of the social cost effectiveness analysis for river basin projects

	Costs		**Benefits**	
Aspect	*Effect/criteria*	*Unit*	*Effect/criteria*	*Unit*
Investment	Costs measures	Euro		
Maintenance	Costs measures	Euro		
Process costs		Euro, time (qualitative)		
Inland shipping	Navigability Compensation Add. dredging Safety	Euro Euro Euro Euro		
Drinking water, groundwater	Compensation	Euro		
Agriculture	Productivity loss	Euro		

	(land, value added)			
Business	Relocation / liquidation	Euro		
Recreation	Relocation / liquidation	Euro	New opportunities	Qualitative
Inhabitants	Relocation Social damage Damage living environment	Euro Destructed houses Euro (compensation)	Improvement living conditions	Qualitative (no. of houses)
Urbanisation	Less flexibility	Qualitative		
Infrastructure	Costs	Euro		
Raw materials	storage	Euro	Extraction	Euro
Polluted soil	Costs	Euro		
Nature	Compensation Nature area	Euro Qualitative	Nature area	Qualitative
Cultural history	Impact Archaeological costs	Qualitative Euro		
Landscape	Impact	Qualitative	Impact	Qualitative
Spatial diversity and coherence	Contextual impact	Qualitative	Contextual impact	Qualitative

In particular in the defining and evaluation of locally desired spatial developments the role of stakeholders has to be expressed. The spatial measures at the local level affect the living environment of civilians most directly. In section 6 we come back on this issue.

Step 1: Local level

To apply the cost effectiveness analysis the following steps have to be taken at the local level:

- Identification of bottlenecks and problems. This identification has to be extensive and not focused only on the impacts of the project Space for the River. For instance, if there is a housing problem as well in the alternative spatial development plan water management measures can be combined with a housing project.
- Identification of the desired spatial developments. Given the example of combining the Space for the River project with a housing problem, one could think of the construction of a bypass river alongside which a new residential area will be developed. This solves the water problem in combination with the creation of an attractive residential living environment. Those desired spatial developments quite often have a locally unique character, given the local context in which the problems occur.
- Definition of the alternatives. Here the focus is on alternative spatial measures (see table 6.3) that could replace the technical measures of the original cost benefit analysis without changing the level of safety.
- After the problem identification and definition of alternatives the impacts have to be identified. The impacts have to be translated in indicators, and by that in measurable units. The impacts (positive and negative) have to be measured for all relevant parties, thus stakeholder identification is part of this phase. Furthermore, the impacts have to be judged on their spatial consequences: is it a generative impact or is it distributive one at a higher spatial level?

Let us explain the approach by the example of the combined development of a bypass river and the development of a residential area. In the first step it is identified that there is a shortage of dwellings and at the same time the dikes of the city are not high enough to comply with the safety standards. In the second step it is identified that an enlargement of the dikes in the city will harm the spatial quality. An alternative would be the construction of a bypass river and an additional advantage will be that this offers an excellent opportunity to develop a highly desired residential area at this new waterfront. Furthermore, the development of this residential area prevents that an alternative location in a nearby forest has to be developed as a residential area. Thus the result of step 3 is the definition of the project and the alternatives. The project is the enlargement of the dike in combination with the development of a residential area in the forest, and the alternative is the construction of a bypass river combined with a residential area at the same location.

The advantages and disadvantages of the alternative over the original project are summarised in table 6.6 and the indicators to measure the impacts in table 6.7.

Table 6.6 Impacts of the alternative project (Source: Renes et al. 2003)

Positive impacts of the project alternative	Negative impacts of the project alternative
Dwelling near the river	No dwelling in the forest
No costs increasing height dikes, etc.	Cost of building islands for houses near bypass
No costs of infrastructure to forest	Costs of bridges to islands
No loss of forest land	Loss of open land near the river
Creation of new nature	Loss of agricultural land
New opportunities for recreation on water	
No harm landscape and cultural history of city due to technical measures on dikes	
No economic loss for trades people due to above mentioned point	

Table 6.7 Indicators to measure the impacts of the project alternative (Source: Renes et al. 2003)

Monetary	Non-monetary
Value of the dwelling	Number of dwellings
Increase of value of dwelling due to location on water front	People on waiting list for a dwelling
	Quality dwellings
	Diversity stock of dwellings
	Number of dwellings on water front

The measurement of this last table will give the politicians the information on which they have to decide to either prefer the original project or the alternative.

Step 2: Regional level

The alternatives at the local level have to be evaluated at the higher regional level to be able to value them on two particular important aspects of spatial quality: diversity and coherence. There might be additional benefits that arise from those alternatives. For instance, the development of a number of nature protection areas might lead to the creation of such a large natural habitat that it is suitable for species that were not identified at the local level (the creation of stepping stones). However, opposite, the comparison of local alternatives at the regional level might also lead to a decrease in benefits. For instance, if a number of cities intend to develop a yacht-basin, the impact of the third and fourth yacht-basin will be less then expected, due to a decrease in demand that was not foreseen at the local level.

Conclusion

To conclude we think that this two step approach offers good opportunities to include spatial quality by a cost effectiveness analysis in a more general cost benefit analysis framework that is entirely based on cost incentives. In the next section we will discuss the - limited - opportunities to include public participation into this combined cost-benefit analysis and cost effectiveness analysis framework.

6.6. INTRODUCING PUBLIC PARTICIPATION INTO THE PROPOSED FRAMEWORK

Cost benefit analysis and public participation

The basic idea behind cost benefit analysis is welfare economics. The aim is to express the change in welfare in monetary units. In principle all costs and benefits of all affected individuals are taken into account, on an aggregate level not on the individual level. Furthermore, cost benefit analysis is only taking monetary effects into account. Non-monetary effects are left out of consideration. In cost-benefit analysis non-monetary effects only are presented in post memory categories and not quantified in monetary costs or benefits.

As a consequence, in practice in cost benefit analysis the actual involvement of individuals concerned is limited. It is possible for the analist to construct a cost benefit calculation from behind his desk, without consulting any stakeholder. Hidden behind the final cost-benefit analysis is a scheme with all distributional effects of cost and benefits between all groups of stakeholders. However, this scheme of distributional effects itself is not presented. Only the aggregate cost-benefit result is presented.

Furthermore, the people concerned are not well known with the method and the reasoning behind it (welfare economics is one of the most complex issues studied in economics). The method is often seen as a 'black box' and by that the results are not taken for granted. To summarise, it is hard to arrive to public support for a cost benefit analysis, since people do not feel attracted to the method.

The main problem is the question how to arrive at a higher level of public acceptance and/or support for cost benefit analysis? Involvement of stakeholders in the decision process in which cost benefit analysis is the main evaluation tool seems to be a major step forward. In our analytical framework we propose a two step approach. The first step concerns a cost benefit analysis based on technical measures to arrive at the defined safety levels. Concerning safety issues, in general measures to increase safety can count on public support. Public resistance against such measures only arises at the local level when sustainability aspects play a role. Therefore, in the second step, we propose cost effectiveness analysis to bring more sustainable elements into the project at the subproject level by spatial measures, without changing the safety levels. In this second step we propose a bottom-up approach, defining alternative - more sustainable - sub projects at the local level and testing them on spatial diversity and coherence at the regional level. In particular at the local level we see opportunities to introduce some elements of public participation in the selection phase of the sub projects. At the local level the public will be most affected by the alternative sub projects. At this level the sub projects directly influence the living environment of the public. Defining and selecting sub projects that can count on strong public support is crucial. Optimisation of the societal support for the scenarios is one of the most important non-binding basic principles of the Space for the River project. Society demands a more open process of spatial planning in which the public is offered the opportunity of consultation. On the other hand, planners want to secure their plans by trying to reach a common consensus. So, openness in the process of policy-making has two advantages: the public has the opportunity to participate in the process and policymakers are well informed and can adjust plans to the desires and interests of the public.

The spider model

In addition to our analytical framework we propose a scenario approach - visualised by so-called spider models. Spider models can be used to define sub projects at the local level based upon extensive communication and interaction between spatial planners and local stakeholders. The proposed approach consists of 6 phases:

Phase 1: Desk research
In this phase an extensive literature survey has to be conducted and the local context has to be analysed to arrive at some first ideas about a selection of more sustainable alternative sub projects at the local level. This phase is used to collect information on possible alternatives.

Phase 2: Stakeholder identification
In the second phase stakeholders are identified, for instance, by advertisements in local and regional newspapers. Stakeholders and interest groups might be Chambers of Commerce, shipping organisations, agricultural organisations, trade unions, environmentalists, tourist agencies, real estate agencies, house owners, et cetera.

Phase 3: Sub project identification
In the third phase a first selection of alternative sub projects is made after a consultation round with representatives of - relevant stakeholders. The stakeholders will provide the project with additional information that was still lacking.

Phase 4: SWOT-analysis
The selected alternatives are analysed on all relevant aspects by a strength-weakness-opportunities-threads analysis. The relevant aspects are already mentioned in table 6.5. In table 6.8 they are categorised by: costs, economy, soil, and sustainability (the categorisation is not unambiguous). Note that spatial diversity and spatial cohesion are not included in table 6.1. The sub projects are evaluated on the aspects spatial diversity and spatial cohesion at the regional level and not the local level

Table 6.8 Categories of relevant aspects

Costs	Soil	Economy	Sustainability
Investment	Drinking water	Inland shipping	Urbanisation
Maintenance	Ground water	Business	Nature
Process	Raw materials	Recreation	Cultural history
Infrastructure	Polluted soil	Inhabitants	Landscape

Spider models are applied to visualise the different alternative sub projects (see figure 6.2). For each alternative sub project a spider has to be constructed in which: a) the present situation, b) the situation after implementing the technical measures, and c) the alternative situation after implementing the sub project, are visualised. As an example they are visualised for a hypothetical situation in figure 6.2. The four categories form the quadrants of the spider model. Each aspect in each quadrant is represented by an axis starting from the interior towards the outer boundary of the spider. The axes are scaled from 1 (interior) to 10 (outer edge). All aspects are scored on this range under the assumption that a higher score represent a better performance. Note that in our example for instance the costs scores are reverse to the actual costs. In the present situation no investments are foreseen (score 10), the technical safety measures of the basic project are relatively cheap (score 5), and spatial measures of the alternative sub-project are rather expensive (score 2).

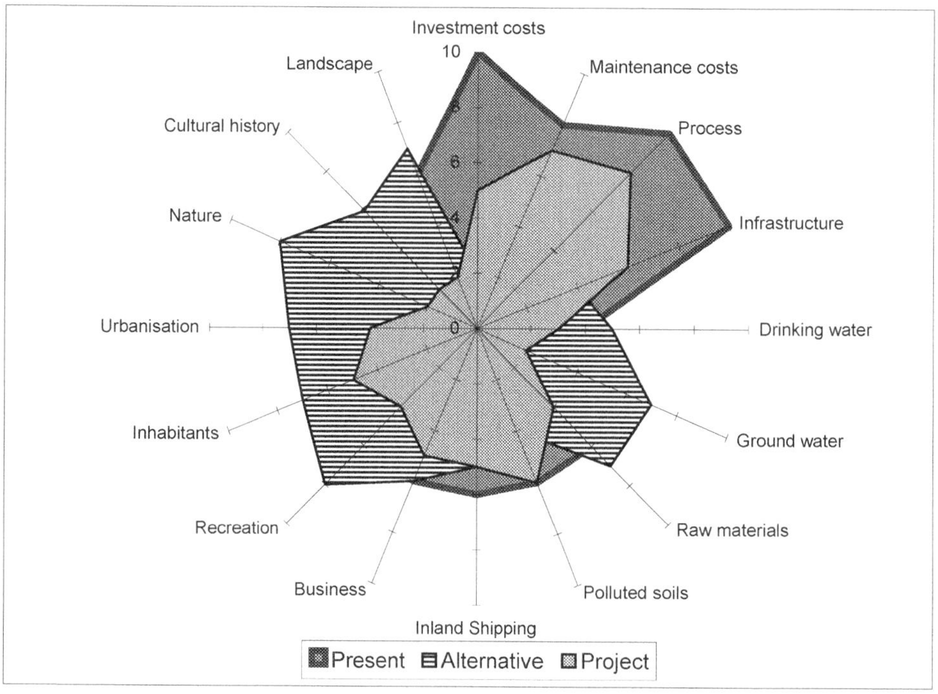

Figure 6.2 Spider model for a more sustainable alternative project

In the spider model there is no weighting between the aspects. In other words: a score of 6 on one aspect is as important as the score of 6 on each other aspect. On the other hand, a score of 6 on one aspect is not the same as a 6 on any of the other aspects. All scores per aspect are connected resulting in a surface representing the integrated mutual relationship between these aspects (see figure 6.2). The advantage of this visualisation by means of spiders is that it is easy to understand how a change in the score of one of the aspects might affect the scores of all other aspects, irrespective to the category the changing aspect belongs to. In this way it can be explained to, for instance, a stakeholder representing environmentalists, that an improvement of the natural environment might have repercussions for level of costs of the sub project or for the inland shipping on the river.

One or more additional spiders can be constructed in which the various sub projects are compared, instead of comparing each sub project with the present situation and the situation after implementing the basic safety project.

Phase 5: Consultation round

The spiders are input for discussion sessions with local stakeholders. By presenting the changes in the present situation caused by the implementation of the basic project based on technical measures and the sub projects based on more sustainable spatial measures in a visualised way, it will be much easier to discuss the pros and cons of each proposal. The consequences of the choice for one of the sub projects can be seen more easily by this visualised way of presentation. In a plenary session the pros and contras of all the (sub) projects have to be explained. After such an introduction all stakeholders are aware of the interests of each of the participants, and each stakeholder becomes aware of the weak spots in his own interest. By this an atmosphere is created where there is mutual understanding of each others interest on the one hand, and the willingness to discuss the relative importance of their own interest in relation to the other - contrasting - interests on the other hand. This might open the doorway to achieve consensus between the stakeholders. After such a plenary session, the discussion can continue in small groups, with stakeholders representing different views in each group. The task is to arrive at consensus within each small group, before presenting a selected sub project by each group in a final plenary discussion. The consultation rounds can be used for the selection and further fine-tuning of the selected sub project. Perhaps a new sub project will be developed from elements of various other sub projects.

Phase 6: Construction of the final sub project

The final goal is to develop a sub project with a solid basis of public support. To arrive at such a sub project it might be necessary to repeat phase 4 and 5 several times. However, in a following phase, the sub project will be evaluated on its spatial diversity and spatial coherence merits at the regional level, together with other local plans. If the sub project doesn't fit in the regional context, the sub project has to be re-examined and falls back in phase 4.

Conclusion

Above, an interesting opportunity is sketched to include public participation into a decision process based on a combined cost-benefit analysis and cost effectiveness analysis framework. Chosen is for an entrance at the local level. At this level people are most directly affected by the implementation of the project. Since in cost benefit analysis the results are presented at an aggregate level, we have chosen to direct public participation towards the phase of the defining and selection of the project. Thus, people are at an early stage confronted with the project and are still in the position to influence the implementation of the project as far as it concerns the local implementation in the people's direct living environment. The overall importance of safety is no subject of disagreement; however, the local fine-tuning may overcome resistance problems major projects in the 1970's had to cope with.

In our application we suggest spider models as a mode to communicate in a more visual way between planners and stakeholders. However, also alternative modes of communication - such as for instance the Flag model approach - will be appropriate.

6.7. SUMMARY AND CONCLUSIONS

The main objective of ADVISOR is to provide an integrated project evaluation framework and methodology for the sustainable governance of Europe's river basins, which will be achieved through the delivery and dissemination of a set of guidelines for integrated river basin project evaluation, consisting of a structured process, criteria and appropriate practical methodological tools.

As part of one of the activities proposed for ADVISOR's Work Package 3, the team of the Free University has developed a combined cost benefit analysis and cost effectiveness analysis framework methodology to evaluate flood protection projects in which public and stakeholder participation is incorporated. The framework is developed for the main Dutch river basin of the rivers Rhine and Meuse.

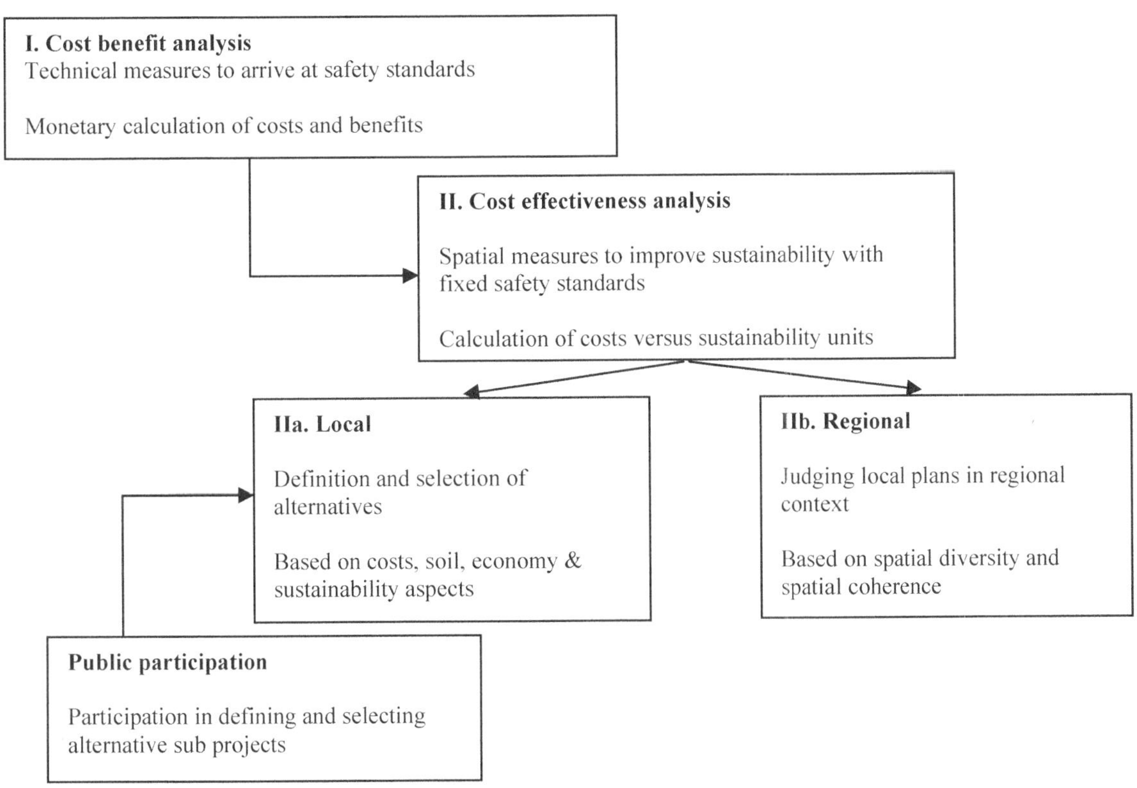

Figure 6.3 Analytical framework

Dike strengthening projects leads to major interventions in landscape and affected the cultural and natural values in the areas concerned. There is clear tension between safety and sustainability aspects. From a cost benefit analysis perspective, technical measures are the cheapest solutions to arrive at the defined safety levels. However, technical measures are in general the least sustainable solutions.

In the evaluation framework we propose, a two-step approach is followed (see figure 7.1). In the first step cost benefit analysis is applied based on technical measures. In general, technical measures, in particular the strengthening of dikes, are the cheapest solutions to arrive at the defined safety levels. For a cost benefit analysis, technical measures offer the best perspectives: the costs of the solutions are the lowest. One might argue that the benefits equal the benefits of more sustainable spatial measures because the additional gains in sustainability of spatial measures cannot be expressed in monetary terms. Nevertheless, from a sustainability perspective, spatial measures are to prefer because they offer more sustainable solutions. The additional gains of those more sustainable measures have to be expressed in other, non-monetary, units. Cost effectiveness analysis offers the opportunity to compare monetary costs (and benefits) with a non-monetary unit.

In the second step a cost effectiveness analysis is applied for more sustainable spatial measures on the level of local sub projects, keeping the level of safety constant. This second step is divided into sub-steps. In the first step spatial quality or sustainability is analysed at the local level. In the second step spatial quality is analysed at the regional level to check for spatial diversity and spatial coherence. In the first step, the impact of desired spatial developments are analysed for four categories of aspects: costs, soil, economy and sustainability. In particular in the first local phase, when alternative more sustainable solutions are defined and selected, we see opportunities to introduce public participation in the evaluation framework. At the local level the public will be most affected by project. At this level the sub projects directly influence the living environment of the people concerned. At this level public support can be created by direct involvement of the public in the development of alternative, more sustainable solutions. On the other hand, public involvement on this level can secure spatial development plans of governments by reaching common consensus. We propose an approach for public consultation in six phases. In our application we suggest spider models to visualise the development of alternative, more sustainable solutions. The spider models can be used as tool to communicate between planners and stakeholders.

Cost benefit analysis, sustainability and public participation are no common combination in scientific research. In cost benefit analysis all pros and cons are measured in monetary terms, whereas sustainability is commonly expressed in non-monetary terms. Cost benefit analysis presents results at an aggregate level, whereas public participation requires an individual design. Nevertheless, we developed an analytical framework in which cost benefit analysis is combined with sustainability by additional cost effectiveness analysis. Furthermore, public participation is included in the analytical framework by the bottom up approach that is developed to define and select alternative, more sustainable solutions. The framework is designed in particular for the evaluation of river basin projects, in which protection against flooding goes hand in hand with the preservation of the typical river landscape.

REFERENCES

De Brucker, K., Verbeker, A. and W. Winkelmans, (1998) Sociaal-Economische Evaluatie van Overheidsinvesteringen in Transportinfrastructuur, Garant, Leuven.

Dunn, W.N. (2003) Public policy analysis, Pearson Prentice Hall, New Jersey.

Hanley, N. (1999) Cost-Benefit analysis of environmental policy and management, In: Bergh, J.C.J.M van den (ed.), Handbook of Environmental and Resource Economics, Edgar Elgar.

Hueting, R.(1980) New Scarcity and Economic Growth, North-Holland, Amsterdam.

Lichfield, N. (1964) Cost-Benefit in Plan Evaluation, The Town Planning Review, Vol. **35**, 160-269.

Ministry of Finance (1995) Kabinetsstandpunt heroverweging disconteringsvoet, Ministerie van Financiën, Den Haag.

Ministry of Transport and Water Works, Ministry of Housing, Spatial Planning and Environment, and Ministry of Agriculture, Nature and Fishery (2002) Startnotitie MER in het kader van de PKB-procedure Ruimte voor de Rivier, Ministerie van Verkeer en Waterstaat, Den Haag.

Munasinghe, M. (1992) Environmental Economics and Valuation in Development Decision-making, Environment working paper no. **51**, Environment department, World Bank, Washington DC.

Nijkamp, P., Ubbels, B.J., and Verhoef E.T. (2002) Transport Appraisal and the Environment, In: D.A. Hensher and K.J. Button (eds.), Handbook of Transport Strategies, Policies and Institutions, Handbooks in Transport 4, Elsevier/Pergamon, Amsterdam.

Van Pelt, M.J.F. (1993) Ecological Sustainability and Project Appraisal. Averbury: Aldershot, UK.

Renes, G., G.M. Bouma, W. Wijnen and Puylaert H.J.M. (2003) Ruimtelijke kwaliteit in de MKBA Ruimte voor de Rivier, TNO Inro rapport 2003-38, TNO Inro, Delft.

Verhoef, E.T. (1996). The Economics of Regulating Road Transport, Edward Elgar, Cheltenham, UK.

Chapter 7

Understanding the Value of Biodiversity in Water Ecosystems: Economics, Ethics and Social Psychology

Clive L. Spash[1, 2], Kevin C. Urama[2], Wendy Kenyon[2], Rob Burton[2], Peter Shannon[2] and Gary Hill[2]

[1] *Geography & Environment Department, University of Aberdeen, U.K.*

[2] *The Macaulay Institute, Craigiebuckler, AB15 8QH, Aberdeen, U.K.*

7.1. INTRODUCTION

The Water Framework Directive (WFD) describes river basin management in terms of achieving "good ecological status". Economic assessment is seen as playing a key role in meeting this goal and in justifying delays; it also can be used to justify a lesser goal of "good ecological potential" where a water body is deemed to be "heavily modified". Such assessment is also required in considering the programmes of measures to be implemented to meet the requirements of the WFD. However, integrated river basin management involves assessment of not only the economic impacts of a project or programme of measures but also the environmental and social impacts. This means going beyond financial appraisal of water management projects and exploring the full range of values associated with environmental change.

Cost benefit analysis (CBA) has developed a set of methods which place a monetary value on improvement or deterioration of the environment, most commonly using the contingent valuation method (CVM); see the project report Spash and Urama (2003). The general aim of the research reported here was to examine the applicability of CBA to quantify environmental change in river basin projects in the context of the WFD. This meant being aware of the extent to which CBA could be used under the WFD to value environmental change and specifically to put monetary values on environmental entities and quality changes in a context relevant to the Directive.

The empirical work involved an extension of the standard CVM to include social psychological drivers and to test their power in explaining willingness to pay (WTP). Investigating the motives that drive responses to CVM surveys was a key concern because of the range of problems which have been encountered in CVM surveys in the past. Specific attention was given to the role of attitudes, norms and ethics in forming and framing a respondent's intention to pay. This approach builds upon previous work addressing attitudes and ethics and their role in understanding how people value the environment and their motives for answering CVM survey questions (Spash and Hanley 1995; Spash 1997; Spash 1998; Spash 2000; Spash 2002). One hypothesis was that CVM is merely a poor measure of attitudes rather than a measure of value, and that alternative means for measuring value, attitudes and ethics could be more relevant in the WFD context. The aims and objectives of the study were to: (i) conduct a state of the art CVM survey that takes account of socio-economic, ethical and social psychology factors; (ii) test the validity of CVM in measuring a population's

 Integrated Evaluation for Sustainable River Basin Governance. Edited by Nuno Videira, Giorgos Kallis, Paula Antunes and Rui Santos. ISBN: 1843391481. Published by IWA Publishing, London, UK.

value for environmental entities; (iii) test the explanatory power of economic, ethical and social psychological drivers of behaviour in a WFD relevant context.

7.1.1. Economics Contributing to the WFD

In favouring one project over a range of others, or determining project specific options to be pursued, many objectives can be relevant including environmental quality, ecosystem integrity, and people's quality of life. In order to meet the accountability requirements expected in modern democracies, budget allocation is meant to be justifiable both in terms of scientific measures (e.g. improved health from drinking water standards) and also public acceptability. For some water pollution problems, such as those involving serious public health or safety concerns, monetary valuation and economic efficiency may be a low priority, but economic analysis may still arise in the form of determining a cost-effective management approach.

Under the WFD there is considerable emphasis on assessing costs. Yet, while there are several articles in the WFD relating to costs the meaning of the term in economics should be remembered. Cost-effectiveness is often considered less problematic than CBA because the range of entities to be assessed is assumed to be far narrower, e.g. supposedly avoiding contentious areas such as valuing human mortality and morbidity or valuing the existence of endangered species. However, costs as a category can cover all such issues. Whether an item is a cost or benefit depends upon the status quo and type of project. For example, the costs of building a dam include habitat destruction which may entail species loss and therefore their valuation as a cost item. Workers are often killed during dam construction meaning the cost of life can become an issue. Once a dam is constructed, the regulation of flows from that dam may be undertaken to improve habitat and as a result a species which was a cost under dam construction become a benefit of a flow regulation project. Thus, cost assessment can require the same approaches (e.g. CVM of species or human life) and be as controversial as benefits estimation (Spash 1997).

The basis of CBA lies in microeconomic welfare theory. Measures of economic value are based on the concept of opportunity cost. That is, when an individual exchanges a good or service they forgo an alternative exchange. How much of all other goods and services an individual is prepared to forgo in order to obtain a specific quantity of a good or service is a measure of their maximum WTP. Similarly, the minimum amount of goods and services an individual accepts in compensation for losing something measures their willingness to accept (WTA). Such an approach to understanding human behaviour is commonly observed in markets but has also been extended to environmental quality. An individual's purchasing power measured in monetary units then determines their ability to command different resources and thereby make effective their choices. The environmental CBA approach has been developed as a practical tool for use in public finance decision support, and can therefore, in theory, be used to evaluate projects being considered under the WFD.

Values associated with a resource, such as a catchment, can be defined in terms of economic value from direct human use (e.g. recreation), indirect human use (e.g. an option for future use, a bequest of assets to future generations, maintaining the existence of species) and non-economic value which includes concerns over the moral standing and ethical status of entities (Spash and Carter 2002). Economic value is one of many possible ways to define and measure value in quantitative units. Although other types of value are often more important, economic value is described by advocates of CBA as providing those in the policy process (e.g. civil servants, government agencies) with a universal numeraire by which comparisons can be made to determine efficient resource allocation.

There are a wide range of goods and services associated with river catchments (e.g., fish, potable water) which may need to be valued under the WFD. Subsets of these are traded in markets so that their economic value could be estimated using supply and demand analysis. Water is also directly used for various commercially productive activities (e.g., hydropower) as an input in the production of goods so that one value measure is the contribution to profits. Valuing other uses of catchments, such as the aesthetics or biodiversity of a wetland ecosystem proves far more difficult. These are, at best, only indirectly bought and sold in commercial markets. In these cases a measure of value (in terms of WTP) can be derived from markets for related goods and services or hypothetical markets. The extended use of CBA to the environment has changed applications from being firmly within a market to a more openly political context. Markets themselves are no longer purely regarded as external entities to be observed but rather institutions which can be designed in hypothetical scenarios.

If these methods are to be useful for decision-making under the WFD then the institution of CBA (specifically CVM) must be seen as valid by individuals who must also be prepared to trade environmental entities for money. They must be aware of and accept this form of indicating to decision-makers their preferences in allocating resources. Evidence from previous studies shows that this is often far from true (Burgess, Clark and Harrison1995, Burgess, Clark and Harrison 1998). Indeed, standard practice in CVM surveys is to exclude from analysis all respondents who refuse to "play the game" and reject the requested trade-off. This may reject up to a third of respondents[14], and thereby taking no account of the opinion of a large proportion of the sample. Thus views which are inconsistent with an economic approach are ignored even though they may be just valid as those of people who do choose to play the game. The standard CVM practice in this area is therefore inappropriate. The research in the current study examines alternative means for addressing respondents' opinions about environmental changes.

7.1.2. The Contingent Valuation Method

CVM has a great appeal among environmental economists due to its claimed ability to measure a wide range of non-priced project impacts from the effects of management options on water quality to changes in biodiversity. The basic technique is to elicit WTP/WTA bids for an identified environmental change through an interview process involving a scenario aimed at simulating a market trade-off. The CVM procedure can be split into six stages (Hanley and Spash 1993): setting up the hypothetical market; obtaining the bids; estimating the mean WTP and/or WTA; estimating the bid curves; aggregating the data, and evaluating the CVM exercise. Many decisions made in designing and implementing the process must be made on the basis of the CVM researchers' own judgment including: (i) the form of the survey (personal interview, mail survey or telephone interview); (ii) the way in which the environmental change is described; (iii) the welfare measures to be used: WTP or WTA; (iv) the 'elicitation' method by which WTP or WTA is measured: direct open-ended question, bidding game, payment cards, or dichotomous choice. A large literature has built up which examines problems associated with CVM methods and makes recommendations on specific aspects of the approach. Much of this literature debates several 'biases' (strategic bias, design bias, information bias) and how to interpret the WTP or WTA bids obtained from the survey (Mitchell and Carson 1989; Hausman 1993;Bateman et al.2002). A selection of contentious issues is discussed in Annex IV of this report. However, a key concern is that respondents should be familiar with the environmental change in question and accept the valuation approach.

7.1.3. The role of Fundamental Ethical Beliefs

The dominant economic theory of decision-making requires a fundamental philosophical assumption; that individuals believe the net utility from the consequences of an action determines whether that action is right or wrong. CBA and its tools, such as CVM, assume that individuals are able and willing to consider trade-offs in relation to the quantity and/or quality of public goods. Debates in environmental ethics have raised the issue of individuals refusing to make these trade-offs and so questions the applicability of economic efficiency arguments (Sagoff 1988; Spash 1993). One aspect of refusal can be a basis of belief in inviolable rights so that actions are intrinsically of value or deontological (Spash 1993; Spash and Simpson 1994; Spash 1997; Spash 2002). Standard CBA rejects such notions because there is an assumed rationality attributed to the ability to make trade-offs, whatever the commodity, as long as enough compensation is offered in return. Empirical work directly addressing the issue has found that a significant proportion of those surveyed treat the environment in a manner which is inconsistent with economic theory (Spash and Hanley 1995; Spash 2000; Spash et al. 2000; Spash 2002).

If an individual believes that certain aspects of the environment, such as biodiversity, have an absolute right to be protected, then that individual is expected to refuse all money trade-offs which degrade the environment. In other words, such individuals should reject the "commodification of environmental resources". Such preferences mean that utility functions including environmental aspects which are to be protected at all cost are undefined for an individual since the axiom of continuity is violated, and that indifference curves collapse to single points, denying the principle of substitution. These preferences are termed lexicographic in the neo-classical economic model because they give absolute priority to one commodity over all others and therefore imply a strict ordering, as in lexicon. A modified form of the proposition can be offered as more likely and, following Spash (1998), was used in the current research in terms of first attaining a minimum standard of living prior to being prepared to defend the environment on the basis of ethical rights.

7.1.4. Learning from Social Psychology: Attitudes and Norms

An area largely ignored by conventional CVM studies is the role of attitudes and norms in influencing WTP. Social psychologists have developed models of human behaviour which challenge the assumptions of mainstream economics but can be integrated into CVM models to test hypotheses about motives for bids. The basic economic model assumes

[14] Recommendation of level of acceptable respondent exclusion by Nick Hanley to DETR during the CVM study on aggregates Department of the Environment Transport and the Regions (1999). The Environmental Costs and Benefits of the Supply of Aggregates: Phase 2. London, Department of the Environment Transport and the Regions: 208.

complete, pre-existing invariant and transitive preferences, which are maximised in static equilibrium frameworks. Conversely, social psychology models focus on the influence of specific psychological attributes (such as attitudes, social/subjective norms, beliefs, perceived behavioural control) on cognitive decision-making processes. Both mainstream economics and social psychology share a focus on the individual and their behaviour. However, the former makes processes of learning endogenous, and excludes social and institutional context. Individual motives are treated as unquestionable and mainstream economics leaves them unquestioned. Social psychology sees behaviour as the result of the interaction between the individual and their psychological, social and institutional environments. The CVM study in this research report introduces factors from social psychology to test the standard economic model of human behaviour.

7.1.4.1. General attitudes: awareness of consequence

Work by Stern et al. has been trying to develop an attitudinal scale to analyse human-environment interactions. In a series of papers they explore such a scale for the prediction of behavioural intentions and develop this around the theme of being aware of the consequences of environmental changes due to human perturbation (Stern, Dietz and Kalof 1993; Guagnano, Dietz and Stern 1994; Stern and Dietz 1994; Stern, Dietz and Guagnano 1995; Stern et al. 1995). This general awareness of consequences (GAC) has been employed to show empirical evidence for at least three centres of concern: self, other humans, and the biosphere. The measures of behavioural intention used include a scale on political action and several variations on WTP questions. However, as reviewed by Spash (2002), the approach to the WTP portion of their work has some inadequacies from the economic viewpoint and, contrary to the stated intention, fails to represent a conventional CVM approach.

Previous work has shown the strong potential of the GAC to explain WTP results (Spash 1998). Factor analysis in that work shows no single centre (e.g. self interest) dominates but rather that mixed motives lie behind WTP responses. This links to concerns expressed that CVM is merely producing a poor attitude measure, and if so then a direct attitude measure using multi-scale items would be far better (Kahneman et al. 1993). However, to our knowledge only one study has so far applied an attitudinal scale to a large public sample while conducting a CVM survey (Spash 1998). The GAC results from that work require replication for increased validity. In addition, as that study recognised, specific attitudes and social/subjective norms may operate differently as explanatory factors and could perform better in explaining WTP. The important question here is whether WTP is related to specific or general attitudes and is therefore more of a behaviour than a general belief.

7.1.4.2. Specific Attitudes and Subjective Norms

The Fishbein-Ajzen (1975) attitude-behaviour model or theory of reasoned action and more recently the related theory of planned behaviour (TPB) of Ajzen (1991) offer some useful insights into how economists could gain by studying social psychology. The TPB model postulates three conceptually dependent determinants of human intention. The first is attitude toward the behaviour (ATB) and refers to the degree to which a person has a favourable or unfavourable evaluation or appraisal of the behaviour in question. The second relates to subjective norms (SN) which refers to the perceived social pressure to perform a specific behaviour. The third is the degree of perceived behavioural control (PBC), which refers to the believed ease or difficulty of performing the behaviour. The PBC is assumed to reflect past experience as well as anticipated impediments and obstacles. As a general rule, the more favourable ATB and SN and the stronger the PBC with respect to a specified behaviour then the stronger should be an individual's intention to perform the behaviour (Ajzen 1991:188). Critical to the success of the model is that it complies with the 'principle of compatibility', i.e. relations between attitudes and behaviour should be assessed (in terms of action, target, context, and time elements) at the same level of generality or specificity (Eagly and Chaiken 1993:164).

In summary, the TPB holds that it is intentions that explain (and can therefore be used to predict) behaviour most directly. ATB is a product of the evaluation of consequences of the act and the probability of the consequence occurring (Ajzen 1991:191). Intention is the cognitive representation of a person's readiness to perform a given behaviour, and it is considered to be the immediate antecedent of behaviour. Attitudes predict behaviour less directly by influencing intentions. Fisbein and Ajzen's model yields three determinants (criteria) of strong attitude-behaviour relationships that the CVM, if designed well, can incorporate: correspondence, proximity, and familiarity. The greater the correspondence between the attitude/intention measures, the more likely the latter will predict the former. Proximity concerns the hypothesis that the fewer are the intervening stages between a component in the model and a behaviour, the greater the predictive power of that component. Finally, familiarity states that the more familiar the behaviour, the more likely the respondent's attitude and/or behavioural intentions will predict that behaviour.

Ajzen, (1991:191) defines ATB as the degree to which performance of the behaviour is positively or negatively valued. Specifically, the outcome's subjective value contributes to the attitude in direct proportion to the strength of the belief, i.e., the subjective probability that the behaviour will produce the outcome in question. This was computed as: $ATB_r \propto \sum b_i e_i$; where "r" is the respondent, "b" the strength of each salient belief, "e" the respondent's subjective evaluation of a belief attribute, and "i" is the specific item (or question) related to the behaviour. For each question eliciting respondent's salient beliefs a corresponding question on subjective evaluation of the specific belief attribute is asked. The product of the response to both questions provides a scale item of ATB, which can be summed to provide an attitude scale.

Subjective norm (SN) is defined as the perceived social pressure to engage in a behaviour. The formulae for SN is: $SN_i \propto \sum n_i m_i$; where "n" is the strength of each normative belief and "m" the respondent's motivation to comply with the referent in question (Ajzen, 1991, p. 195). An SN scale is computed as directly proportional to the sum of the product of n and m.

A final control which determines intention and action is a set that deals with the presence or absence of requisite resources and opportunities. "The more resources and opportunities individuals believe they possess, and the fewer the obstacles or impediments they anticipate, the greater should be their control over the behavior" (1991:196). Beliefs about resources and opportunities underlie PBC just as beliefs about consequences of a behaviour determine ATB, and normative beliefs determine SN. PBC is directly proportional to the sum of products of each control belief "c" and the perceived power "p" of the particular control factor to facilitate or inhibit performance of the behaviour: $PBC_i \propto \sum c_i p_i$.

7.1.5. Developing a New Model of Environmental Valuation

In the light of the preceding discussion, this study extended the conventional CVM survey to include ethical variables, social psychology drivers and indicators of environmental awareness. Based upon previous work which included ethics and attitudes (Spash 1998; Spash 2000; Spash 2000) a new model was expected to improve the ability to explain WTP results. CVM practitioners have argued that WTP models that achieve an explanatory power of 15% are acceptable (Mitchell and Carson 1989; Bateman et al. 2002), which leaves a considerable amount of unexplained variation.

The various elements required in the new model are show in Figure 7.1. The elements to be tested in this research project are shown as coloured. The green elements are those taken from the TPB. Socio-economic, ethical and general awareness of consequences are factors all previously tested in the context of CVM. The relationship between elements is unclear but can be hypothesised as follows.

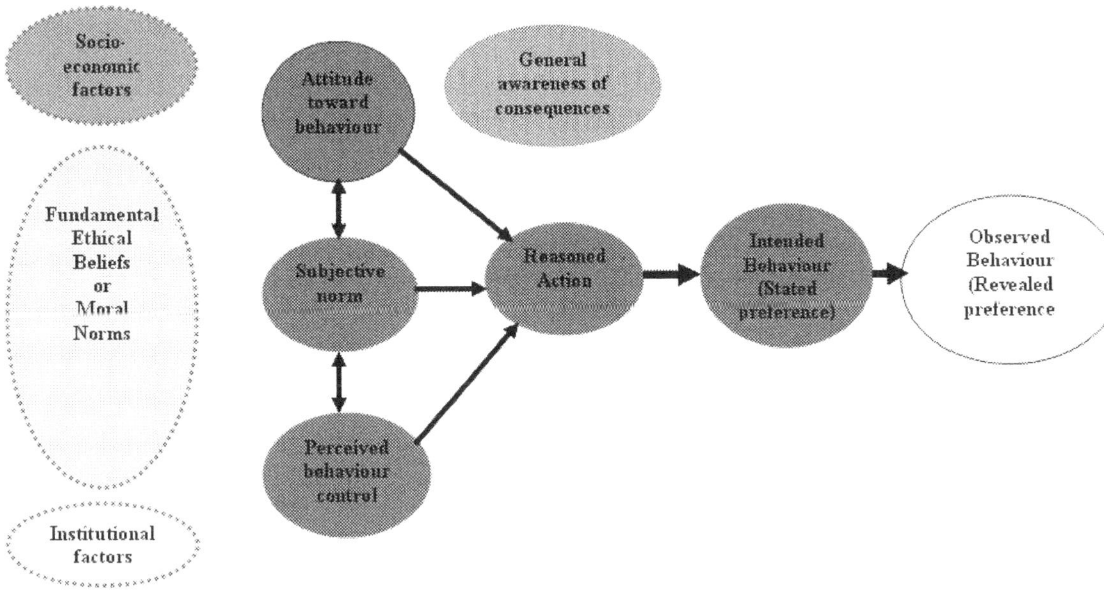

Figure 7.1 New conceptual model for environmental valuation

Institutional factors are implicit in the PBC an individual believes they have and also as is implicit within subjective norms. Institutional factors also set the context for any valuation exercise. Under the CVM choices are made as to the form of payment, to whom a payment is made and so on. Issues such as trust arise and the extent to which people believe they control a situation is important. An historical understanding of any problem is necessary to comprehend the institutional setting. Institutions can also be regarded as informal arrangements existing between individuals or groups. In this regard social norms are relevant. In the TPB subjective norms are measured and can be viewed as addressing the existence and relevance of informal institutions or behavioural rules.

Socio-economic factors will influence specific attitudes and subjective norms most directly. Fundamental ethical beliefs could lie behind the TPB model or operate as another factor along side social norms. The GAC is a general attitude measure and as such might relate to specific attitudes or replace them with respect to a given behaviour and thus directly relate to reasoned action. Rather than clutter the diagram with these speculative relationships there are no lines drawn but rather the research agenda is left open and the aim of this project was to inform about the potential relationships and their strength.

7.2. SURVEY DESIGN

Survey design requires many decisions by the analyst and is an art. A key issue is the selection of the case study and the institutional setting. The CVM requires a realistic scenario in which people might be asked their WTP or WTA for an environmental change. While the trade-off is hypothetical it is meant to be one which people can easily imagine occurring in practice. The survey normally takes 20-30 minutes, which restricts the ability to explain complex problems. Thus, different techniques for conveying information are required and language must be kept simple. As a result of such issues pre-testing is particularly important and has been expanded in recent years to include focus groups. The focus group is used to review use of information aids (e.g. pictures, graphics, show cards), wording, and general acceptance of the scenario(s) posed. In the following section the case study choice is explained. This is followed by a discussion of specific issues in the survey design process. Finally pre-test findings are reported. The aim is to show how changes to the survey occurred and how the final design was achieved.

7.2.1. Case Study Selection

In order to test the model, described in Introduction, a suitable case study had to be selected. A number of issues were relevant in identifying an appropriate site. First, the case study had to be relevant to the WFD context of the ADVISOR project. In particular, cases were identified that were relevant to Article 4 of the Directive which sets out when economic arguments are required to justify the designation of water bodies as "heavily modified" and therefore subject to a reduced ecological standard. The economic justification required in such circumstances provided a suitable context for the CVM study. Second, the scenario and change to the scenario should be feasible, practical and believable. For example one case considered would have required valuation of the hypothetical decommissioning of a major coal fired power station, Longannet, on the Forth estuary which would have provided respondents with an unrealistic scenario. Third, potential case studies needed to be relatively simple and limit the number of complex issues being addressed. This helps make the valuation task manageable for the respondents. Many sites involve multiple complex issues such as climate change, conservation of biodiversity, diffuse pollution and water quality issues. Fourth, a meaningful and appropriate payment vehicle is required which matches the case study context. For instance, one potential case study involved improving water quality for migratory salmon. The payment vehicle might have involved either higher prices for fish or a licence fee for anglers. However, these instruments failed to address the specific issues involved in the valuation scenario. Fifth, design is greatly aided if there is readily available relevant and accurate information on the environmental change being proposed. This may be scientific literature or government reports which included impacts and how they might be mitigated. This information allows specification of the marginal change upon which the economic welfare measure is meant to be based.

Following consideration of a number of potential sites and examination of the issues discussed above, the case study selected for the survey was the Tummel catchment in Scotland which is depicted in Figure 7.2. The location is in the Grampian Highlands of Scotland, and drains into the River Tay covering an area of 1713 km^2. The area includes eight reservoirs and pre-existing lakes which are used for hydro-power generation, along with the rivers and streams draining into and connecting them. The catchment is one of two cases in the UK considered for potential derogations due to hydro power schemes at the time of the study (the other was on the River Dee). The river Tummel is part of a freshwater system made up of an intricate network of rivers, streams and lakes with 1,253 km of river channel and 77 km^2 of standing waters. The Tummel and its major tributary, the River Ericht, drain the southern and western parts of the catchment whilst the River Garry and its main tributary, the River Tilt, drains the north eastern part. The Garry joins the Tummel just upstream of the town of Pitlochry. The Tummel area has a very low population density (less than 0.10 persons per hectare) and the area is classified as a less remote rural zone (Black et al. 2002).

At the time of the study, the Tummel catchment was provisionally designated as a heavily modified water body under the WFD and a number of implementation measures were being discussed to achieve good ecological status or good ecological potential in the river catchment. Black et al (2002) had identified measures which would address some of the requirements of the WFD in the catchment. One of these was the introduction of a compensation flow regime from the dammed lochs to mimic the natural flow in some of the rivers within the catchment; the aim being to restore the diversity and abundance of species and habitats in the river catchment. The work by Black et al (2002) provided significant amounts of relevant, accurate and accessible information around which to develop the CVM questionnaire. The survey scenario was based upon information contained in the report concerning physical characteristics of the water bodies, land and water use in the area, human impacts, ecological status and possible mitigation actions.

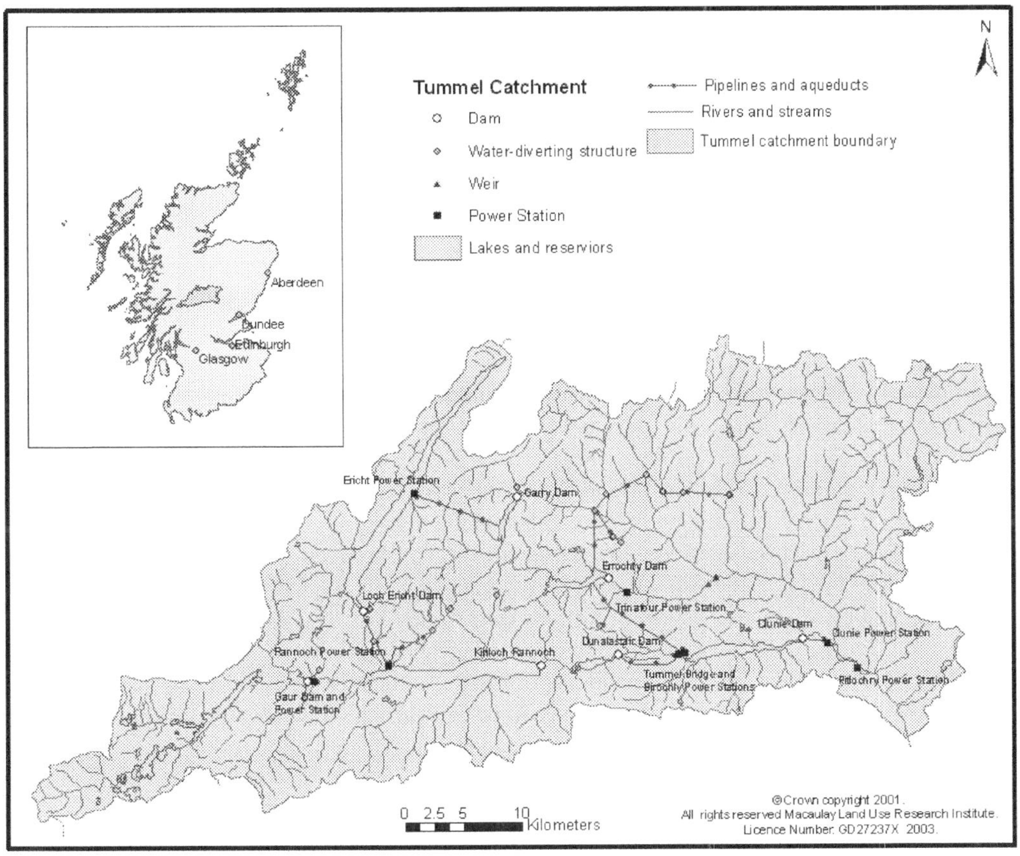

Figure 7.2 Location of the Tummel Catchment

Black et al. (2002) identified workable policy options that required consideration by managers dealing with implementation of the WFD in the Tummel catchment. Grounded in real policy concerns, the Tummel provided a scenario of reduced biodiversity in the rivers downstream of the hydro-dam due to low flow. This was a suitable basis for the development of a practical, feasible and believable scenario for a WTP question. This also offered a relatively simple scenario where a single issue (increasing biodiversity) could be isolated and explained easily.

The scenario also presented a suitable payment vehicle in the form of increased energy bills. Increasing river flows from the hydro-system would potentially reduce electricity generation and increase costs for the hydro-power companies. Such costs would then be (wholly or partially) transferred to electricity consumers. Eliciting WTP for increments in household electricity bills was therefore seen as a realistic bid vehicle suiting the scenario.

7.2.2. The Process of Survey Design

The CVM survey followed a specific design developed over several years and studies (Spash 1993; Spash and Hanley 1995; Spash 1997; Spash, Aldred and Rutherford 1997; Spash 1998; Spash 1998; Spash et al.1998). This had five sections (A to E): introductory framing and information; monetary valuation and follow-up; environmental ethics, attitudes, SN and PBC; socio-economic data; interviewee and interviewer evaluation.

Section A, asked the respondent about a range of public policy issues such as education, the environment and the NHS, in order to determine their level of concern related to each issue. These introductory questions make the respondent think about the theme of the questionnaire within the wider context of other policy issues. The results provide information on the relative importance to the respondents of the different issues while making explicit the context of biodiversity improvement as only one area of potential expenditure. Such framing has been suggested to avoid embedding problems. Following the introductory question, the respondent was asked to focus on biodiversity and about their knowledge of biodiversity, both in general and related to water bodies. The final questions in Section A related to the case study on the River Tummel. Respondents were asked whether they were familiar with the area and then provided with information about the water courses, the hydropower schemes and informed about the associated changes in biodiversity.

Information for the survey was selected and reviewed through several stages which involved: reports on the case study and associated environmental impacts; expert consultation with aquatic ecologist; social psychologist; economists; end users through informal discussion with RSPB, SEPA, SNIFFER, Environment Agency, DEFRA; the general public focus group discussion; researchers via presentations to project partners; the general public via a pilot study. Three types of information were used in the final survey. First, information about biodiversity in general, aquatic biodiversity and the web of life provided background on the declines in biodiversity due to hydro-power, and the potential increases in biodiversity as a consequence of compensations flows. Secondly, information about the Tummel catchment, the hydro-power schemes in the area along with details of the decline in biodiversity specific to those schemes and thirdly information that familiarised respondents with the two scenarios. The "business as usual" scenario described biodiversity at 14% of natural levels due to reduced water flow in the catchment. The alternative scenario involved an increase in water flows and an associated increase in biodiversity in the catchment to 70% of natural levels at the expense of reduced energy generation and increased electricity bills. A follow-up question asked respondents how well (7-point scale) they had understood the impacts described.

Section B comprised WTP and open-ended follow-up questions to explore the reasons for responses. Three types of valuation questions were asked. First, the payment in principle question addressed whether respondents found acceptable the idea of paying an increase in their energy bill to improve biodiversity without reference to the Tummel. Second, the specific WTP question where the 14% to 70% change in biodiversity in the Tummel area is specified as the object of value. Thirdly, a qualitative question on how valuable the restoration of biodiversity in the River Tummel was to the respondent. The latter was included to explore the consistency between qualitative and quantitative preference expression.

The WTP question can be asked in a variety of formats (see Annex IV). The open-ended format was selected as the most easily understood by respondents and as avoiding leading respondents to a desired answer (as occurs with some other methods due to starting point bias and anchoring effects). As already mentioned, the payment vehicle was the electricity bill. A realistic story could be told as to why the electricity bill would increase in order to increase biodiversity. The mechanism was also regarded as reasonably neutral although some respondents might react negatively to the company or lack trust in their carrying out the action. Follow-up questions were used to allow these effects to be measured.

Section C related to the extended CVM model discussed above and comprised questions on ethical beliefs, the TPB and attitudes. The idea behind this section was to obtain measures of the underlying motives for respondents WTP. This can prove a cognitively challenging task for respondents, although past work, upon which this study was based, has shown the vast majority of respondents are able to answer without finding the task too complicated (Spash 1998). The aim in the design stage was to refine the initially large set of questions down to a minimum while maintaining the integrity of the model.

Section D covered relevant socio-economic data including gender, age, education and income. The categories used corresponded with the 2001 Census. Finally, Section E asked for interviewer evaluation of the difficulties encountered by the respondents and whether other people bedsides the respondent were present when the interview was conducted. This information proves useful for validating survey results and identifying any problems which specific individuals encountered.

7.2.3. Questions for the new model

Section C starts with ethical questions to measure the level of belief in species protection extending from animal rights to economic to human centred. The aim of the questions in this section was to identify non-economic ethical positions so that they could be analysed for influence on stated preferences. The method used replicated exactly Spash (1998). The respondents were asked to state the extent to which they saw rights as relevant to the Tummel case study. Those making a specific attribution of rights were then probed further because a general discontent with trade-offs may disappear upon the specification of extreme consequences. Thus respondents were asked to reflect upon the extent to which their refusal to trade was absolute by considering a potential conflict with their own standard of living. Specifically, the respondents were told that "A major aim of maintaining natural flows of water is to provide habitat for endangered wildlife species such as otters and kingfishers". They were then asked which one of four statements most closely matched their opinion about the scheme to get hydro-power companies to release more water to the rivers to mimic a natural flow. The respondents also had a fifth choice "can't answer – this is too complicated". Rights based respondents were then given an additional question which split them into strong and weak categories.

Section C2 covered the TPB questions. A total of thirteen pairs of questions were asked to elicit respondent's ATB for WTP to restore biodiversity in the Tummel catchment. An example of the ATB pairing is: "Paying more for electricity to restore biodiversity will increase the diversity and abundance of plant and animal species in the Tummel area." This was evaluated from extremely likely to extremely unlikely. This was combined with the subjective evaluation of that specific belief: "Increasing the diversity and abundance of plant and animal species in the Tummel area is…" with the evaluation being from extremely bad to extremely good. Six pairs of SN questions were included. An example which elicits a normative belief is: "My spouse / partner would think that I … pay more for electricity to preserve biodiversity in the Tummel area.", where the evaluation was from "should" to "should not". This is combined with a motivation to comply assessed by asking: "Generally speaking, how much do you want to do what your spouse/partner thinks you should do?" with the scale going from "not at all" to "very much". A total of five PBC questions were included as single items (i.e., not paired). An example is: "I can *easily* afford to pay more for my electricity." with responses being from "strongly agree" to "strongly disagree". All evaluative scales were 7 point with 4 designated as "cannot decide".

7.2.4. Testing the Design

Development of the survey took place through two focus groups of 10 individuals each and a pilot survey. A total of 81 questionnaires were completed by face-to-face personal interviews for the pre-test.

7.2.4.1. Focus Groups

The focus groups were selected by quota sampling to cover a spectrum of the target population. One of the focus groups was held in Edinburgh (urban) and the other in Blairgowrie (rural). Both were led by a trained facilitator. The lessons learned from the focus groups were used to revise the survey before the pre-test. The aims of the focus groups were to: test respondents understanding of each question; assess the information provided on the biodiversity impacts of hydro-power systems in the Tummel catchment; develop the TPB scales. The trained facilitator went through each question with the focus group participants, discussing wording, sequencing, style and eliciting participant's opinion regarding any changes that would help them to better understand the questions. The participants preferred the use of simple and familiar words in the impact descriptions and framing of the questions. For example, the use of words like 'areas surrounding a river' and 'variety of species' were preferred to technical words like 'river catchments' and 'species richness' respectively. The information provided in the pre-test survey used pictures, graphics and texts developed in collaboration with a water ecologist. Respondents preferred the graphic representation and actual pictures as opposed to computer generated images which were presented as an option. The computer images had been tested because they allowed a more precise manipulation of the scenario.

Some questions in Section D were thought to be intrusive. This was particularly the case for information on respondent's income. Despite being regarded as a standard predictor of WTP in CVM studies, this question is often problematic with large item non-response and inconsistent (i.e., compared to job categories) and unbelievable stated amounts (i.e., very low). A mixture of category first then specific income request was developed for the main survey as an attempt to improve response.

Respondents found the questions in Section C on TPB cognitively demanding. As a result the wording of questions and descriptions were changed to reflect the preferred simple language. The number of questions on TPB in the pre-test was also larger than intended for the final survey as the aim was to select those which performed best.

7.2.4.2 Pre-test survey

The pre-test survey was sub-contracted to an accredited survey company (Feedback Market Research Ltd.) and took place between 12th July and 13th August 2003. The sample was a quota spread on age, gender, rural-urban location, income and standard occupational classifications. Interviews were conducted in several Scottish council areas (see Annex V). There were slightly more female than male respondents, an even age spread across categories, over sampling in low-income groups and under sampling in low education categories. The survey took an average of 16 minutes 55 seconds (min. 10 max. 45 minutes).

About 68% of the pre-test sample had no understanding of what biodiversity meant before the survey, and only 2.5% said that they understood the term very well. The majority (57%) had above average understanding of the biodiversity impacts described in the survey while 16% failed to understand the impacts at all. The majority of respondents clearly believed in a human cause-effect relationship being responsible for biodiversity change.

Responses to being WTP in principle showed 47% in favour, 42% against, 1% refused and 10% "don't know". The specific WTP responses had a large percentage of "refusals"/"don't knows" at 53% and 12% respectively while 33% of the responses were positive and only 2% zero bids. This suggests that about 14% of those who were WTP in principle were unwilling to do so for the specific restoration project. The pre-test results raised some concerns over the payment vehicle as many non-payers were concerned over high electricity prices. The mean WTP from 58 respondents was £3.22 with both minimum and media being zero and maximum £15 from. Those responding "No" to WTP in principle were coded as zero bids.

Results for the ethical categorisation showed a quarter of the sample took a strong species rights position while the weak position was relatively small. The positions consistent with economic theory amounted to just under 40%. A relatively high proportion found the question too difficult to answer (just over 17%).

The pre-test was used to reduce (by statistical analysis) the number of questions for estimating the TPB model because of the cognitive demands of the task. The aim was also to test whether a relatively short set of survey questions could be successful in measuring the TPB, which would normally be the subject of a dedicated interview process. In the pre-test, there were some strong and statistically significant relationships between elements of the model. Although the aim was not for use of the results they at least showed the model could operate in a survey context and even with a small sample gave good results (e.g. Chronbach's alpha 0.9).

Table 7.1 Pre-test Survey: Level of difficulty in answering questions

	1	2	3	4	5	6	7	Total
Section A	6	1	0	0	2	1	1	11
Section B	3	2	3	0	5	0	1	11
Section C	0	0	6	2	3	0	3	14
Section D	4	4	0	0	2	0	1	11
Degree of seriousness	0	1	1	4	6	10	29	49

Notes: Difficulty scales 1 Not at all to 7 Extremely difficult; Seriousness scale 1 Not at all serious to 7 Very serious

Table 7.1 reports the evaluation of the questionnaire by the interviewees. The table shows that Section C (i.e. the ethical/TPB section) was the most difficult for the respondents to answer. More than half of the respondents took the survey very seriously while about 91% of those who responded to the evaluation questions ranked their seriousness at 5 or above.

7.3. MAIN SURVEY BASIC RESULTS

7.3.1. Sample and Descriptive Statistics

A probability proportional to size sampling technique was used with multi-stage quota sampling by location, gender, age, income and standard occupational classifications. The Scottish Census 2001 served as the sampling frame. The aim was to get a good representation of each strata of the Scottish population and ensure ease of comparison of data between the different social groups. The sample consisted of 1,012 face-to-face interviews that took place between August and December 2003. A map and detail of the geographic coverage can be found in the questionnaire, which makes comparisons with the Scottish population in urban and rural council areas. Division into urban and rural was by population density per hectare. Statistically insignificant difference existed between the sample and Census population by council area.

There was a relatively high percentage of under 25 year olds for a CVM survey as the aim was not to match the population i.e. with children, but to target adults or heads of household. Housewife/husband, the retired and to a lesser extent the unemployed were over represented. This is a normal occurrence for in-home interviews as these groups tend to be more readily available. While the retired are over represented the over 55 years group is under represented and this reflects a failure to obtain interviews from the employed elder part of the population. This shows itself also in the under representation of professional/managerial classes. Again this is a normal occurrence with the higher income categories proving difficult to interview e.g., working later hours, having a high opportunity cost of time. Income data is discussed again in Section 4, but can be noted here to suffer a high non-response rate. Educational attainment is hard to compare directly with Census data categories. An approximate comparison for Scotland is that 19% have a first degree, higher degree, or professional qualification which equates to 21% from categories 4 and 5 in the survey, while in the Census 33% are classified as no qualifications or none that are recognised which would be category 1 and part of 2 in the survey or 51% for both categories together. Annex V gives detail on how the sample compares with Census data in terms of gender, age groups, and standard occupational classes.

7.3.2. Public Perception of Water Ecosystem Biodiversity

7.3.2.1. Biodiversity Compared to Other Public Policy Issues

At the start questions were used to frame the survey in terms of public policy issues. These asked for respondents concern about seven issues. There were four general issues raised which were the quality of education, the quality of public transport, the state of the National Health Service and the level of criminal activity. These were followed by three environmental issues: the quality of the global environment, biodiversity in Scotland and the quality of the natural environment within respondent's locality. More specifically, the question on biodiversity was phrased to ask concern about: "The variety of living things (plants and animals) in Scotland". Concern about crime received the highest frequency of "very concerned" responses, whilst concern about public transport received the least. The particular focus of this work is related to the biodiversity issue via water ecosystems and superficially there does appear to be some difference between biodiversity and other issues. However, statistical testing for difference between the frequency distributions of each public policy issue and "concern for biodiversity" proved insignificant.

7.3.2.2. Knowledge and Opinions on Biodiversity and Species Loss

The extent of public understanding of the concept of biodiversity was revealed by the next stage of the framing. Respondents were asked to rate how well they thought they understood the term "biodiversity" (7 point scale). This question was asked prior to any information being given which defined biodiversity or described the web of life. Under 8% of the sample felt they understood the term "very well" and more than half of those questioned had no understanding of the term at all while 70% had below average understanding[15].

This question was followed by a definition of biodiversity in terms of species, genetic and ecosystem elements. Then a graphic representation was used to explain the "web of life" in a river catchment with specific reference to the river Tummel being introduced. The food chain and interdependence of different species from fungi and algae to otters and birds was explained. Following on from this information delivery respondents were asked two questions on their opinions about the status of water species biodiversity. First respondents were asked to give their opinion on the trend in number of species in rivers, lochs and estuaries in Scotland over the last 10 years. Second they were asked to attribute responsibility for this trend to human actions or natural processes or both. The results are cross-tabulated in Table 7.2. There were significant relationships between opinions about species change and their cause. There was a clear majority who believe species in Scottish water systems had declined over the last decade. Humans were seen as primarily responsible for a decrease in species numbers while a minority believed in stability seeing humans and natural processes as responsible. Natural processes were generally regarded as being dominated by human change with only 2% believing they were the main reason for species trends. While most people were able to give both an opinion on species trend and the cause, there was a substantial minority (17%) who stated they "did not know" an answer in both instances.

7.3.2.3. Tummel Catchment and Biodiversity Loss

Before being given any specific information on the River Tummel catchment respondents were asked how well they knew it. The majority of respondents had poor knowledge of the area with three quarters scoring below average. The implication for valuation is that use values for any benefits derived from an environmental improvement would be a low priority.

Next respondents were given information about the extent of hydropower schemes in the Tummel area (i.e. eight hydropower stations, nine main dams and a system of pipelines and aqueducts). A range of impacts were described due to the hydro schemes, e.g. reduced flows down-stream, turbine injuries to animals, flooding of habitat up-stream, and reduced diversity and abundance of insect species in specific lochs by 86 percent. Before showing pictures of these impacts respondents were asked how well they understood what they had been told. There was above average understanding by 60% of the sample with almost a quarter having very good understanding. However there was also a significant minority with poor understanding. This information is important for analysis of the WTP results reported in Section 7.4 and can be used as an independent variable in regression analysis.

[15] This result is similar to that found in a poll by MORI where 63% of a UK sample did not know what the words 'biological diversity' meant. MORI (1998) cited in Christie, M., Hanley, N; Warren, J., Hyde, T., Murphy, K., and Wright, R. (2004) "Developing measures for valuing changes in biodiversity on farmland using choice experiments and contingent valuation". A paper presented at the Agricultural Economics Society conference, Imperial College, London, 4 April 2004.

Table 7.2 Trends in Water Species Abundance

		Who is responsible for this species trend?				N
		Mainly human actions	Mainly natural processes	Both human actions & natural processes	Don't know	
Decreased	Actual	383.0	12.0	246.0	6.0	647
	Expected	278.6	14.1	217.0	137.4	
	Adj. Residual	(13.9)	(-1.0)	(4.0)	(-21.1)	
Increased	Actual	24.0	5.0	20.0	3.0	52
	Expected	22.4	1.1	17.4	11	
	Adj. Residual	(0.5)	(3.8)	(0.8)	(-2.8)	
Stable	Actual	13.0	5.0	46.0	32.0	96
	Expected	41.3	2.1	32.2	20.4	
	Adj. Residual	(-6.1)	(2.1)	(3.1)	(3.0)	
Don't Know	Actual	14.0	0.0	26	173	213
	Expected	91.7	4.6	71.4	45.2	
	Adj. Residual	(-12.1)	(-2.5)	(-7.4)	(24.1)	
Total	N	434	22	338	214	1008
	%	43	2	34	21	100

Chi-Square Tests

	Value	df	Asymp. Sig. (2-sided)
Pearson Chi-Square	684.938(a)	9	0.000
Likelihood Ratio	679.460	9	0.000
Linear-by-Linear Association	402.503	1	0.000

Note, no cells should have an expected frequency less than 1, and the minimum expected count here is 1.13. Less than 20% of the cells should have an actual frequency below five and here 3 cells (18.8%) have an expected count less than 5. Sample 1008 valid cases i.e. 4 refusals.

7.3.3. Valuing Water Biodiversity and Ecosystem Structure

In order to assess the value of restoring species diversity and ecosystem structure in the Tummel three types of questions were asked:

(1) Payment in Principle: respondent's agreement with being willing to accept an increase in their energy bill in order to increase biodiversity;
(2) Specific WTP: the maximum additional amount they would be willing to pay each quarter on their electricity bill over the next year to restore biodiversity in the river Tummel and its surrounding area from 14% to 70%.;
(3) Qualitative Value: the personal value of the restoration of biodiversity in the river Tummel and its surrounding area from 14% to 70%. This was measured on a scale from 1 to 7 (not at all to extremely valuable).

The first two questions followed each other directly and were preceded by the scenario on improving the Tummel River in stream flows by reducing electricity generation to mimic a more natural flow regime making the electricity generating companies incur extra costs. The third question came after the follow-up questions on reasons for payment and refusals to pay. The responses to the first two questions are reported in Table 7.3.

7.3.3.1. Payment in Principle

The idea of asking a payment in principle question has been advocated to reduce protest responses and allow a separation between the general scenario of paying for an environmental improvement and the specific detail of an improvement. About 38% of the whole sample was willing to accept an increase in their energy bills to restore biodiversity. As stated above, those who accepted payment in principle were then asked to state an actual amount of money they were WTP each quarter. The combined proportion of the refusals, zero responses and don't knows was large (72%).

The largest single category was the zero bid group (43%) which resulted entirely from stating "NO" to the payment in principle question i.e., nobody gave an explicit zero value to the specific scenario after having agreed to pay in principle. However, about 10% of those who accept payment in principle then crossed over to the "don't know" category when faced with the specific WTP question for the restoration proposal.

Table 7.3 WTP Category of Response

		N	Percent
Payment in Principle Rejected			
	Zero Principle	437	43.2
	Don't know Principle	180	17.8
	Refuse Principle	8	0.8
Payment in Principle Accepted			
	Positive WTP	282	27.9
	Zero Specific	0	0.0
	Don't know Specific	104	10.3
	Refuse Specific	1	0.1
Total		1012	100.0

As noted by Spash (1998), the treatment of these response categories poses a critical problem in the analysis of monetary values obtained from conventional CBA methods. Large numbers of respondents may be treated arbitrarily as protest bidders and excluded from analysis, or as holding no value for an environmental change when they in fact do. The reasons and motives that underlie responses therefore need careful attention and are explored below.

7.3.3.2. WTP for the Biodiversity Improvement

Only the respondents who answered "YES" to the WTP in Principle question were asked the specific WTP question. Clearly, in this case, this removed all those who would have been classified as zero bidders under a CVM survey which had no such prior question. Those who stated they were unwilling to pay in principle are therefore treated as zero bidders (43% of the sample).

Table 7.4 shows the average WTP has a high standard deviation and was £5.62 each quarter for the restoration of biodiversity by 56% in the Tummel catchment and its surrounding area. The median was zero which indicates that over 50% of the sample was unwilling to increase their electricity bills. In this case, using the mean figure as an indication of public WTP for biodiversity restoration in the Tummel could be misleading and aggregation to the general public extremely problematic. The median is suggested as an indicator of political willingness to accept a general payment being imposed e.g. a compulsory fee. If all electricity bills were increased this would fail to meet the political criteria and could be expected to lead to protests. At the same time the 28% of positive bidders cannot be taken as the percentage of those who value the environmental improvement. An unknown proportion of other respondents are also likely to hold a positive (including non-monetary) value. In order to gain some understanding of the potential population holding positive values the follow-up questions to the bids can be used.

Table 7.4 WTP for Biodiversity Improvement in the Tummel Catchment

	Min	Max	Median	Mean	Std Dev	N
WTP positive bids only	£1.00	£120.00	£10.00	£14.33	£17.33	282
WTP with zero bids	£0.00	£120.00	£0.00	£5.62	£12.90	719

Notes: excludes 9 Refusals, and 283 Don't Knows.

Table 7.5 presents the reasons for non-payment. The most heavily cited set of reasons related to the payment vehicle. The pre-test had shown some sign of this problem with the bid vehicle. The responses in the main survey were consistent and similar in the three categories identified: the largest group objecting to electricity company pricing polices and profits with the belief evident that they could pay for an environmental change themselves, the second being a related issue of distrust but relating to the expectation that the money would not go to the project specified, the last was a more practical problem of the respondent not being the one who pays the electricity bill. Payment vehicle problems together accounted for around 38% of reasons for non-payment. The implication with the first two reasons is that they are protests against the companies and have noting to do with the actual environmental improvement, i.e. if there were another way in which the same improvement could be achieved and funded then these individuals might well pay. The next largest category of responses are a common set which concern lacking funds with either a claim of income constraint (often suspect or hard to substantiate give lack of income responses) or other priorities. These two reasons are taken as valid zero bids in terms of standard economic models. A desire for further information is also common and unsurprising given the stated lack of understanding by some respondents. The related desire for costing is an aspect of the open-ended payment mechanism which might have been removed by using a payment card but theory suggests costs should be avoided as the aim is to assess benefits, i.e. WTP. The desire for government to pay is often linked to comments about taxation being paid already and suggested these individuals would also protest against a tax bid vehicle. Two reasons more specific to this case study are the desire for a solution which replaces hydropower dams by alternative (environmentally friendly) energy sources, and the giving of positive bid reasons while making no bid.

Positive bids prove more difficult to clearly classify as signified by the relatively large proportion of those under "other". However, there are some clear reasons given which can be grouped together. The first two referred to in Table 7.6 appear similar but in fact differ in generality. The concern to protect Nature or the environment is non specific and appears to signify a general contribution to a good cause rather than the specific improvement. The concern for wildlife is more specifically focussed on helping non-human species and for some is an animal welfare issue. The worry about species extinction covers reasons where this is obviously a key concern. The desire to preserve species/the environment for future generations is clear but often stated without reference to the specific scenario. Indeed the specific scenario or aspects of it are mentioned by very few respondents as the main reason for their bid. The reference to wanting to "do our bit" raises questions over the extent to which this is seen as a moral obligation because of the damages which humans have created more generally. These individuals seem to be appealing to the duty everyone has to contribute to improve environmental quality. The extent to which such inferences are valid is explored in greater detail using the ethical variables in Section 7.4.

Table 7.5 Reasons for Non-payment[1]

	N	%
Payment vehicle: Don't pay the bill	50	6.8
Payment vehicle: Electricity bills/profits too high already	176	24.1
Payment vehicle: Distrust companies	50	6.8
Too poor/can't afford	137	18.8
Change unimportant/other priorities	76	10.4
Lack information	52	7.1
How much would it cost?	44	6.0
Government/others should pay	46	6.3
Try alternative energy sources	26	3.6
Support but no bid	19	2.6
Other	22	3.0
Don't know	32	4.4
Total	730	100.0

Notes: [1] Excludes the 282 positive bids

Table 7.6 Reason for Positive Specific WTP

	N	%
Help protect environment/Nature	81	28.7
Help wildlife (animals, species)	60	21.3
Worried about species extinction	11	3.9
For the future (children, grandchildren)	26	9.2
Feel have responsibility ("do our bit")	34	12.1
Specific scenario	15	5.3
Other	55	19.5
Total	282	100

7.3.3.3. Qualitative Valuation of the Biodiversity Improvement

In order to provide a comparison with the specific WTP results respondents were asked to rate the value of the 56% restoration of biodiversity in the Tummel catchment and its surrounding area on a 7 point scale. The results show 23 people who gave a positive WTP in fact rated the proposed improvement below 4 on the scale. This is consistent with a desire to contribute to help the environment and wildlife in general and most of these (WTP but low value) respondents are grouped in these categories. More significant is the number of non-payers who actually do claim to value the biodiversity change. Indeed 63% of the non-payers ranked their quantitative valuation at 4 or above. However, some of these appear to be inconsistent rather than just protest bidders who are masking a positive WTP. Some people ranked the change as extremely valuable while giving their non-payment reason as seeing the change as unimportant or below other priorities. Many of the payment vehicle protesters did state they valued the proposed change in biodiversity as did those in "lack information" and "cost unknown" protest response categories.

7.4. TESTING THE NEW MODEL OF ENVIRONMENTAL VALUATION

As discussed in Section 7.1, there are key elements of environmental valuation which appear to be missing from economic explanations of human behaviour. In particular economics assumes a preference utilitarian philosophy so that consequences determine whether an action is right or wrong. A major alternative and common philosophy is rights based. The prevalence of such alternatives can be tested within the CVM context and this work conducted such a test. The psychological model in mainstream economics assumes a narrow self interest which ignores the role of attitudes and social rules. Some branches of institutional economics do address concerns over social rules as informal institutional arrangements. The aim here was to include both attitudes and social rules. More specifically the TPB was adopted to allow an expanded test of the motives behind behavioural intentions such as WTP. This model is explored here and compared with a more standard socio-economic model and an ethical model. This section shows how different factors can be measured with implications for the explanation of WTP for biodiversity within the context of water management. The first two sections concentrate on ethical and TPB models respectively. This is followed by regression analysis of various models.

7.4.1. Ethical Motives for Behaviour

The classification of ethical positions on a rights versus consequentialists basis was done with regards to endangered species protection with respect to the scenario of hydro-power companies changing in stream flows. The picture which emerges is similar to past research in that a third or more of the sample find rights based statements most closely match their opinions about the scheme to protect endangered species such as otters and kingfishers. About half the sample take a consequentialist position consistent with the philosophy underlying economic theory. In terms of the relative strengths of the two rights based positions this sample shows more people maintain a strong rights position than found in previous work (Spash 2000). This might indicate an increasing polarisation in the environmental debate.

A long standing but largely untested hypothesis in the CVM literature was that individuals protest against CVM and bid nothing rather than take part in a process which implicitly buys and sells improvements in biodiversity. That is, if you hold a philosophical position which disagree with monetary valuation of the environment you will refuse to cooperate with a survey which requests such a valuation. In the current work approximately 43% of those who bid positively did so on the basis of non-economic reasoning (i.e. they failed to choose consequentialist ethical categories to support their reason for giving). Past work has found even higher percentages (e.g. Spash 2000 found 53%). In fact, most respondents try to be accommodating to the requests made of them. So we find rights based individuals do bid positively as well as appearing in other bid categories (zero, don't know, refuse). However, the question is whether there is a significant difference due to an individual's ethical position or are these distributions merely random.

Table 7.7 shows that ethical categories do indeed have a strong influence on WTP. The chi-squared results are highly significant. This also shows some deviation from what might be expected on the basis of standard economic explanations. The strong rights and consequentialist favouring species seem close together in that they are over represented in positive bid categories and underrepresented in the zero bid category. Consequentialist favouring people and humans first are also similar to each other in showing the opposite bias, i.e. less likely to bid positively and more likely to be zero. Those placing humans first also tend to be more likely to reject the scenario in principle while those with strong rights and favouring species tend to accept the idea of payment in principle. The weak rights give relatively weak results and are far closer to an expected frequency distribution across categories. As Table 7.8 shows those in the strong rights position also have the highest mean WTP and, with the consequentialist favouring humans, the highest median WTP.

Table 7.7 Ethical Position on Species Protection Compared with WTP Category

		WTP Category				Total
		Positive WTP	Zero Principle	Don't know Principle	Don't know Specific	
Strong species rights	Actual	104.0	66.0	25.0	34.0	229
	Expected	74.5	97.0	32.5	25.0	229
	Adj. Res.	4.9	-4.9	-1.7	2.2	
Weak species rights	Actual	13	28.0	9.0	5.0	55
	Expected	17.9	23.3	7.8	6.0	55
	Adj. Res.	-1.5	1.3	0.5	-0.4	
Consequentialist favouring species	Actual	122	79.0	38.0	39.0	278
	Expected	90.4	117.8	39.5	30.3	278
	Adj. Res.	4.9	-5.7	-0.3	2.0	
Consequentialist favouring people	Actual	36	131.0	29.0	13.0	209
	Expected	67.9	88.6	29.7	22.8	209
	Adj. Res.	-5.4	6.8	-0.2	-2.5	
Humans first	Actual	2	57.0	20.0	2.0	81
	Expected	26.3	34.3	11.5	8.8	81
	Adj. Res.	-6.1	5.4	2.8	-2.6	
Total N		277	361	121	93	852

Chi-Square Tests	Value	df	Asymp. Sig. (2-sided)
Pearson Chi-Square	144.266(a)	12	.000
Likelihood Ratio	160.460	12	.000
Linear-by-Linear Association	18.000	1	.000

Notes:

1. No cells have an expected count less than 5 (minimum expected count is 6.00).
2. Excludes refusal of WTP and "don't know" responses to ethical categories: missing N=160

Table 7.8 Mean and Median WTP by Ethical Position

	Strong species rights	Weak species rights	Consequentialist favouring species	Consequentialist favouring people	Humans first
Mean	£10.6	£4.4	£7.3	£2.5	£0.5
5% Trimmed Mean	£7.2	£2.3	£5.8	£1.2	£0.0
Median	£5.0	£0.0	£5.0	£0.0	£0.0
Maximum	£120.0	£50.0	£60.0	£50.0	£30.0
Minimum	£0.0	£0.0	£0.0	£0.0	£0.0

7.4.2. TPB Model Results

The TPB model has three key elements: ATB, SN and PCB. In the current context the intended behaviour is the WTP for the change of in-stream flows in the Tummel catchment to improve biodiversity. The method chosen to estimate these three factors required a list of forty three questions. There were thirteen paired questions on ATB with each pair consisting of one question on the strength of belief that a behaviour will have a specific outcome (likely/unlikely) and a second on the desirability of that behavioural outcome (good/bad). Similarly, for the SN there were six paired questions with the first addressing the salience of the norm in terms of what others think should be done about a specified behavioural act and the second the motivation to comply with those others. The subjective norms related to six social relationships with the following individuals/groups: spouse/partner, colleagues, children, parents, friends and environmental groups. Each paired question is intended for use as a product to be summed into the respective ATB or SN scale.

The PCB questions were slightly different. The recommend approach under TPB is again to use the sum of products. In this case the first is the individuals belief in their control over a given factor (e.g. paying money) and the second is their perception of the factor in achieving a given end (e.g. biodiversity improvement). The type of behaviour under a CVM survey was more difficult to conceptualise as a product and space precluded a large range of additional questions. The aim was to assess the extent to which respondents perceived that they could perform the target behaviour at will. In this respect there are in fact two target behaviours involved: first is control over payment, second is achieving a biodiversity improvement via that payment. Other studies were found to have used single questions rather than products and given the complexity of the CVM context this approach was taken. Two questions then addressed the first, and three the second target behaviours; this lead to a more experimental approach, which makes the results here of particular interest.

The first question in each pair for ATB and SN was phrased negatively and the second positively to encourage cognitive engagement. The PCB questions were a mixture of positive and negative phrasing. The results give a measure from 1 to 49 for the ATB and SN scales and a scale from 1 to 7 for the PCB. That is, the summed products (ATB, SN) and summed items (PCB) are divided by the number of items in the scale. Item non-response leads to exclusion in the summation process[16]. Generally, the more favourable ATB and SN and the stronger the PBC with respect to a specified behaviour then the stronger should be an individual's intention to perform that behaviour.

The first stage in data analysis of the results requires testing the integrity of the expected scales. This was done using reliability and factor analysis. The ATB and SN scales have good initial reliability including all items. ATB has a Chronbach's alpha of 0.9416 from 976 responses, and SN has 0.8374 from 789 responses. The integrity of the SN scale could not be improved by item removal but that for ATB was increased to 0.9466 by removing one item (ATB10 sum product of questions C2.19 and C2.20). This item related to the price increase in the electricity bill and the desirability of using less electricity as a result. From the protest bid reasons there was a clear concern over the electricity bill and this question was as a result unclear with regard to the association with the biodiversity improvement scenario. A second item (ATB4 sum product of questions C2.7 and C2.8), which was of concern in this regard, related to biodiversity improvements leading to tourism, which could be seen as either beneficial or problematic. In the factor analysis this was a weaker element but still made a statistically significant contribution while ATB10 failed to do so.

Scale reliability for PBC was much weaker. The Chronbach's alpha for all five items was 0.4245 from 983 responses. This was increased to 0.4896 (from 988 responses) by removing the two items relating to electricity bill payment. Testing those two items as a separate scale shows a weak result with alpha 0.2745 (from 988 responses). However, factor analysis supported the separation of the scale as reported next.

Factor analysis gave four factors. The first two were ATB and SN scales. The third and fourth were the two elements of the PBC scale. The item ATB10 was confirmed as the weakest of the ATB scale and in an unrotated component matrix fell marginally below the 0.4 loading criteria. As the item is borderline and reduce alpha reliability it was dropped from the regression analysis. As noted above PBC was expected to be more difficult and here splits into two factors. These are weaker but at the same time do hold together statistically and are in-line with the item focus with factor 3 being biodiversity related and factor 4 electricity company related.

Analysis of the partial correlation coefficients showed ATB and SN to have a consistently significant relationship to the valuation measure, although the relationship is weaker for qualitative valuation. The two factors are also significantly related to each other and this correlation is increased under qualitative valuation. Overall the relationship between internal elements of the TPB seems to be one where ATB is closer to SN and PBC relates more to SN than ATB.

PBC has a mixed strength and significance. The separation of the two aspects of PBC shows they are uncorrelated themselves and have different relationships with other variables. Their main consistent relationship is with SN. In terms of payment versus qualitative scenarios of valuation, PBC with respect to biodiversity improvement (PBCbio) is most significant for the qualitative value (although unexpectedly negative and a small correlation) whereas PBC with respect to electricity bills (PBCelec) is significant in the two other case and with a stronger correlation. The change in significance of PBC between the payment in principle and specific WTP stages indicates those who "refused" or "didn't know" at the first stage also lacked a belief in their ability to control the behaviour being requested, i.e. biodiversity improvement via electricity company bills.

---

[16] Note using a mean summation would result in respondents who answer say one set of paired ATB questions being treated exactly the same as one who answered all 13 pairs. This would be highly misleading, implying that all scale items are identical to the one for which an answer was obtained.

7.4.3. WTP Analysis

This section examines the variation in WTP to increase biodiversity in the Tummel catchment according to socio-economic characteristics, ethical categories and the TPB model (using factors ATB, SN and PBC). The statistical relationships are explored using standard multivariate regression analysis of models in socio-economics, ethics and social psychology which are then combined. The bid function for WTP was explored using a log linear form where zero bids were assigned the value of £0.001 to obtain the natural log (lnWTP).

Of the 1012 individuals interviewed there were 282 positive bids and 437 zero bids; see Section 7.3. The 293 refusals and don't knows are excluded from analysis under this section. Regression analysis further reduces the 719 sample size due to item non-responses. In this respect income is a particularly problematic variable. In this study, 38% of the sample (382 individuals) refused to give their income by category while 68% (688 respondents) refused to answer when asked their disposable income. Comparing income data with occupation class implies some inconsistency. Including the variable would reduce the total sample size considerably. Instead, following Spash (1998), a dummy variable INCREF was created to cover respondent refusal to give income level. In addition, dummies were used for occupation classes and education both of which can act as a surrogate for income. The variables to be discussed in the various models are described in Table 7.9 and the coefficients for all models are summarised in Table 7.10.

Table 7.9 Model Variable Descriptive Statistics

	N	Min	Max	Mean
Urban Area	1012	0	1	0.64
Quality of global environment	1011	1	7	5.15
Water system biodiversity in Scotland has increased	1012	0	1	0.64
Water system biodiversity in Scotland has remained stable	1012	0	1	0.1
Understanding of information on impacts	1003	1	7	4.56
EDUNIV	996	0	1	0.16
EDUNIVPG	996	0	1	0.05
Managment, Professional	978	0	1	0.16
Serious	1004	1	7	5.69
Income Refused	1012	0	1	0.68
Strong species rights	1012	0	1	0.23
Weak species rights	1012	0	1	0.05
Consequentiailist favouring animals	1012	0	1	0.27
Consequentialist favouring humans	1012	0	1	0.21
ATB Scale with 12 items (ATB10 removed)	967	2.5	49	23.68
SN Products Sum Scale	789	1	49	15.99
PBC Sum Scale on Electricity Company	988	1	7	3.80
PBC Sum Scale on Biodiversity	988	1	7	2.92
LNWTP	719	- 6.91	4.79	- 3.32
Valid N (listwise)	523			

7.4.3.1. Socio-economic Model

The socio-economic model shows the largest contribution to explaining variation in WTP is from the respondents understanding of the information on the impacts of hydropower generation on biodiversity in the Tummel catchment. This is followed by the belief in an upward trend in biodiversity; both are significant and positive. In the later case a negative correlation might be expected as a belief that biodiversity is increasing could be taken to signify that all is well. The explanation may be that although biodiversity is on the increase it has a long way to go. Alternatively, if people are more aware of species and notice their presence they may also value them more and want to see more of them and protect them from harm. The lowest contributing variable is also concerned with belief in the trend in Scottish biodiversity and in this case concerns stability which may be most easily understood as signifying recognition of a need to pay to increase biodiversity. Another significant environmental variable is the concern for the global environment which, given reasons for positive bids is unsurprising but does imply more of a general motive than consideration of the specific issue being discussed. The model also includes a variable on how seriously the respondent took the survey in the opinion of the interviewer and this proves positive and significant. Thus, those who were cognitively engaged had a higher WTP.

Table 7.10 Regression Models Compared

	Socioecon. Model	Ethics Model	TPB Model	Combined Model
(Constant)	-10.653 (0.000)	-6.448 (0.000)	-12.469 (0.000)	-17.897 (0.000)
Urban Area	1.368 (0.000)			2.806 (0.005)
Quality of global environment	0.219 (0.024)			0.998 (0.319)
Water system biodiversity in Scotland has increased	2.029 (0.000)			3.157 (0.002)
Water system biodiversity in Scotland has remained the same	1.250 (0.034)			1.667 (0.096)
Understanding of information on impacts	0.429 (0.000)			-0.441 (0.659)
EDUNIV	0.996 (0.029)			1.959 (0.051)
EDUNIVPG	2.282 (0.006)			3.046 (0.002)
Management, Professional	0.908 (0.045)			0.685 (0.494)
Serious	0.390 (0.001)			0.722 (0.471)
Income Refused	-1.096 (0.001)			0.008 (0.994)
Strong species rights		5.204 (0.000)		6.231 (0.000)
Weak species rights		2.422 (0.001)		0.731 (0.465)
Consequentiailist favouring animals		5.057 (0.000)		5.577 (0.000)
Consequentialist favouring humans		1.473 (0.001)		1.245 (0.214)
ATB Scale			0.172 (0.000)	9.633 (0.000)
SN Scale			0.124 (0.001)	4.745 (0.000)
PBC Scale Electricity			0.600 (0.000)	4.862 (0.000)
PBC Scale Biodiversity			0.230 (0.038)	3.349 (0.001)
Adj R^2	0.223	0.225	0.475	0.569
F-test	20.134 0.000	52.992 0.000	125.732 0.000	39.319 0.000
Total DF	667	718	552	522

Education was important and increased WTP, but was also partially correlated with occupation classes as included by the management/professional dummy variable. The two University education categories of undergraduate and post graduate degrees reduced the significance of the occupation variable. The occupation class and education variables covered the absence of income data. This is also addressed in a different way by the income refused variable which was negative and significant showing that refusal to give income was associated with reduced WTP. This may be interpreted as those refusing to pay wishing to avoid being seen as miserly and so refusing to admit their income level. Being from an urban area increased WTP and could signify higher incomes levels but may also be due to general attitudinal differences (this indicates a need for further analysis of urban-rural differences).

Overall the model shows the higher are education and occupation levels the greater WTP. While being concerned for the environment, understanding biodiversity impacts and taking the survey seriously are also key factors. The model's adjusted R^2 at 22% is quite respectable for a CVM study.

7.4.3.2. Ethical Model

While all the ethical variables are significant the greatest contribution to explaining WTP comes from the strong rights position. The four variables together explain as much as the ten socio-economic variables. The importance of the strong rights factor shows how WTP is based upon non-economic motives. However, the standard economic approach describes ethical positions as leading to protests while here we see they contribute to a positive WTP. The problem for the standard approach is that these results offer little comfort for regarding the resulting WTP as reflecting a trade price or valuation of the biodiversity in terms of a preference utilitarian ethical theory.

7.4.3.3. Social Psychology Model

The ATB scale contribute most under the TPB model followed by the SN and PCB electricity scales which by comparison with some other regressions also add well to the explanatory power. The PCB biodiversity is much weaker and far less significant. This is despite the scale being the more robust of the two on the reliability tests and in factor analysis. The adjusted R^2 at 0.48 is very high within the CVM context where 0.14 to 0.20 is more normal and taken as acceptable. Clearly the TPB model is highly relevant to the WTP results and this has strong implications for their interpretation.

7.4.4. Model Result Comparison

A reduce form of the models used above is shown in this section which increases the explanatory power with regard to WTP. In the cross comparison of the different models in Table 7.10 the results are shown for a model combining all factors. The adjusted R^2 is 0.57 which is very high for a CVM model. However, there are many variables which are insignificant and this combined model shows that a reduced form with very few variables and improved statistical significance can be created. Indeed the socioeconomic variables normally associated with a bid curve appear to be the least important.

The approach for the reduced form model is to look for variables which remain significant at the 5% level while contributing to the overall explanation of WTP. This is shown in Tables 7.11, 7.12 and 7.13. The top five variables relate to the TPB and ethical models allowing 53.6% of the variation in WTP to be explained. The remaining five variables explain just an additional 3.1% of variation. Thus, the key to understanding WTP results lies in the social psychology and ethical beliefs of the respondents. In terms of the TPB model elements all three are present and PBC actually dominates SN in terms of explanatory power. For the ethical variables the strong rights position comes through as might be expected, but also the consequential favouring animals. Overall the model has intuitive explanation and appears statistically robust.

Table 7.11 Contribution of Variables to a Combined Model

Model	Variables Entered	R	R^2	Adj R^2	SE
1	ATB Scale	0.609	0.371	0.370	3.576
2	PBC Scale Electricity Company	0.663	0.439	0.437	3.379
3	SN Scale	0.690	0.477	0.474	3.268
4	Strong species rights	0.705	0.497	0.493	3.206
5	Consequentiailist favouring animals	0.735	0.540	0.536	3.068
6	Urban Area	0.741	0.550	0.544	3.040
7	EDUNIVPG	0.747	0.557	0.552	3.017
8	Water system biodiversity in Scotland has increased	0.751	0.563	0.557	2.998
9	PBC Scale Biodiversity	0.756	0.572	0.564	2.973
10	EDUNIV	0.759	0.575	0.567	2.963

Table 7.12 Summary Statistics

R	R Square	Adj R Square	Std. Error	F	F Sig
0.759	0.575	0.567	2.963	72.107	0.000

Table 7.13 Variables in the Combined Model

	B	SE B	Beta	t	Sig t
(Constant)	-13.74	0.50		-27.49	0.000
ATB Scale	0.14	0.01	0.36	10.45	0.000
PBC Scale Electricity Company	0.51	0.09	0.18	5.76	0.000
SN Scale	0.09	0.02	0.17	4.87	0.000
Strong species rights	2.52	0.34	0.24	7.53	0.000
Consequentiailist favouring animals	2.19	0.32	0.22	6.85	0.000
Urban Area	0.73	0.28	0.08	2.61	0.009
EDUNIVPG	2.39	0.70	0.10	3.41	0.001
Water system biodiversity in Scotland has increased	0.84	0.28	0.09	2.97	0.003
PBC Scale Biodiversity	0.353	0.105	0.102	3.36	0.001
EDUNIV	0.825	0.373	0.065	2.21	0.027

7.5. CONCLUSIONS

The use of economic valuation techniques appears potentially very strong within the WFD. As a result CVM is likely to play a key role in determining monetary values of environmental change. There are several issues which environmental valuation research over the last decade has raised which, if addressed, could aid those wishing to assess environmental values for policy purposes. More specifically the importance of motives behind values (including ethical positions, environmental attitudes and norms) can be seen as offering greater insight into how policy should be designed. Ethical categories and scales from social psychology were used in this study to explore their validity both as explanations for responses to CVM surveys and as stand alone measures of human motives. The former use allowed the validity of WTP results to be tested by investigating the consistency of motives with those assumed by the standard economic model. The latter use may prove more effective than CBA approaches especially where divergent and contested values are likely to arise.

The CVM study reported here used the TPB model in order to develop measures of attitudes and norms regarding biodiversity and to test whether they could explain WTP. The results of the focus groups and pre-test survey gave confidence in the questionnaire design. The level of difficulty reported by the respondents especially with regard to Section C was expected. This was partially addressed by the fact that the TPB model used in the main survey was a cut down version of the pretest (removing 14 less statistically significant TPB questions).

Overall, the analysis found that human valuation of biodiversity improvement is significantly influenced by both ethical categories and the social psychological variables of the TPB. More specifically, attitudes significantly influenced responses to all types of valuation question with subjective norms coming close behind. PCB seems to be more valuation mode specific. The results both reconfirmed the role and importance of ethical positions and showed how standard economic models are flawed. The implication for the WFD is that multiple values will be relevant to management decisions and monetary valuation will fail to address public concerns. In addition CVM is seen as reflecting psychological and ethical factors rather than producing a trade price for environmental change or an economic welfare measure.

REFERENCES

Ajzen, I. (1991) The theory of planned behaviour. *Organisational Behaviour and Human Decision Processes* **50**, 179-211.

Bateman, I. J., Carson, R. T., Day, B., Hanemann, M., Hanley, N., Hett, T., Jones-Lee, M., Loomes, G., Mourato, S., Ozdemiroglu, E., Pearce, D. W., Sugden R., and Swanson, J. Eds. (2002) *Economic Valuation with Stated Preference Techniques: A Manual.* Cheltenham, Edward Elgar.

Black, A. R., Bragg, O. M., R. W. Duck, A. M. Findlay, N. D. Hanley, S. M. Morrocco, A. D. Reeves and J. S. Rowan (2002), Heavily Modified Waters in Europe: Case Study on the River Tummel. Geography Department, University of Dundee and Department of Economics, University of Glasgow.

Burgess, J., Clark, J. and Harrison, C. (1995) *Valuing Nature: What Lies Behind Responses to Contingent Valuation Surveys?* University College London.

Burgess, J., Clark, J. and Harrison, C. M. (1998) Respondents' evaluations of a CV Survey: a Case Study Based on an Economic Valuation of the Wildlife Enhancement Scheme, Pevensey levels in East Sussex. *Area* **30**(1), 19-27.

Department of the Environment Transport and the Regions (1999) The Environmental Costs and Benefits of the Supply of Aggregates: Phase 2., Department of the Environment Transport and the Regions: 208, London

Eagly, A. H. and Chaiken, S. (1993) *The psychology of attitudes.* Fort Worth, Texas, Harcourt Brace Jovanovich.

Fishbein, M. and Ajzen, I. (1975) *Belief, Attitude, Intention and Behavior: An Introduction to Theory and Research.* Reading, MA, Addison-Wesley.

Guagnano, G. A., Dietz, T. and Stern, P. C. (1994) Willingness to Pay for Public Goods: A Test of the Contribution Model. *Psychological Science* **5**(6), 411-415.

Hanley, N. and Spash, C. L. (1993) *Cost-Benefit Analysis and the Environment.* Aldershot, Edward Elgar, England.

Hausman, J. A., Ed. (1993) *Contingent Valuation: A Critical Assessment.* Amsterdam, North-Holland.

Kahneman, D., Ritov, I., Jacowitz K. E. and Grant, P. (1993) Stated willingness to pay for public goods: A psychological perspective. *Psychological Science* **4**(5), 310-315.

Mitchell, R. C. and Carson R. T. (1989) *Using Surveys to Value Public Goods: The Contingent Valuation Method.* Resources for the Future, Washington, D C.

Sagoff, M. (1988) *The Economy of the Earth: Philosophy, Law, and the Environment.*, Cambridge University Press, Cambridge.

Spash, C. L. (1993) Economics, ethics, and long-term environmental damages. *Environmental Ethics* **15**(2), 117-132.

Spash, C. L. (1993) Estimating the Importance of Inviolable Rights: The Case of Long-Term Damages and Future Generations, Scottish Economic Society.

Spash, C. L. (1993) Intergenerational Transfers and Long Term Environmental Damages: Compensation of Future Generations for Global Climate Change due to the Greenhouse Effect. *Department of Economics*. Laramie, University of Wyoming**:** 378.

Spash, C. L. (1997) Environmental management without environmental valuation? *Valuing Nature? Economics, Ethics and Environment.* J. Foster. Routledge**,** 170-185, London.

Spash, C. L. (1997) Ethics and environmental attitudes with implications for economic valuation. *Journal of Environmental Management* **50**(4), 403-416.

Spash, C. L. (1998) Contingent valuation. *The Elgar Companion to Consumer Research and Economic Psychology*. P. Earl and S. Kemp. Cheltenham, UK/Northampton, MA, Edward Elgar Publishing Ltd**:** 128-134, USA.

Spash, C. L. (1998) Environmental Values and Wetland Ecosystems: CVM, Ethics and Attitudes. Department of Land Economy, University of Cambridge**:** 111.

Spash, C. L. (1998) Investigating Individual Motives for Environmental Action: Lexicographic Preferences, Beliefs and Attitudes. *Ecological Sustainability and Integrity: Concepts and Approaches*. J. Lemons, L. Westra and R. Goodland. Dordrecht, The Netherlands, Kluwer Academic Publishers. **13,** 46-62.

Spash, C. L. (2000) Ecosystems, Contingent Valuation and Ethics: The case of Wetlands Re-Creation. *Ecological Economics* **34**(2), 195-215.

Spash, C. L. (2000) Ethical Motives and Charitable Contributions in Contingent Valuation: Empirical Evidence from Social Psychology and Economics. *Environmental Values* **9**(4), 453-479.

Spash, C. L. (2000) Multiple Value Expression in Contingent Valuation: Economics and Ethics. *Environmental Science & Technology* **34**(8), 1433-1438.

Spash, C. L. (2002) Empirical Signs of Ethical Concern in Economic Valuation of the Environment. *Economics, Ethics, and Environmental Policy: Contested Choices*. D. W. Bromley and J. Paavola, Blackwell**:** 205-221, Oxford.

Spash, C. L. (2002) *Greenhouse Economics: Value and Ethics*. London, Routledge.

Spash, C. L. (2002) Informing and Forming Preferences in Environmental Valuation: Coral Reef Biodiversity. *Journal of Economic Psychology* **23**(5), 665-687.

Spash, C. L., Aldred, J. and Rutherford, C. (1997) Fenlands Valuation: Pretest CV Survey and Initial Analysis. Cambridge Research for the Environment.

Spash, C. L. and Carter, C. (2002) Environmental Valuation Methods in Rural Resource Management. *Nature and Agricultural Policy in the European Union: New Perspectives on Policies that Shape the European Countryside*. F. Brouwer and J. van der Straaten. Cheltenham, Edward Elgar Publishing Ltd**:** 88-114.

Spash, C. L. and Hanley, N. (1995) Preferences, Information and Biodiversity Preservation. *Ecological Economics* **12**(3),191-208.

Spash, C. L. and Simpson, I. A. (1994) Utilitarian and Rights-Based Alternatives for Protecting Sites of Special Scientific Interest. *Journal of Agricultural Economics* **45**(1), 15-26.

Spash, C. L. and Urama, K. C. (2003) Monetary Valuation Methods Reviewed for Relevance to River Basin Management: A Report for the EC Project Integrated Evaluation for Sustainable River Basin Governance (ADVISOR). Geography and Environment Department, University of Aberdeen**:** 16.

Spash, C. L., J. van der Werff ten Bosch, S., Westmacott and Ruitenbeek J. (1998) Lexicographic Preferences and the Contingent Valuation of Coral Reef Biodiversity in Curaçao and Jamaica; Report to The World Bank. Delft, Resource Analysis**:** 135.

Stern, P. C. and Dietz, T. (1994) The Value Basis of Environmental Concern. *Journal of Social Issues* **50**(3), 65-84.

Stern, P. C., Dietz, T. and Guagnano, G. A. (1995) The New Ecological Paradigm in Social-Psychological Context. *Environment and Behavior* **27**(6), 723-743.

Stern, P. C., Dietz, T. and Kalof, L. (1993) Value Orientation, Gender and Environmental Concern. *Environment and Behavior* **25**(3), 322-348.

Stern, P. C., Dietz, T., Kalof, L. and. Guagnano, G. A (1995) Values, Beliefs and Pro-Environmental Action: Attitude Formation Toward Emergent Attitude Objects, *Journal of Applied Social Psychology* **25**(18), 1611-1636.

Chapter 8

Scenario Workshops as a Tool for Sustainable Water Resource Governance: The Case of the Island of Naxos, Greece

Dionisia Hatzilacou, Giorgos Kallis, Harry Coccosis, Eleni Svoronou, Alexandra Mexa, Stefanos Xenarios and Ilias Tziritis

Environmental Planning Laboratory, University of the Aegean, Greece

This chapter describes the experimental application of a "scenario workshop" as a tool for the participatory formulation of an action plan for the management of water resources in Naxos, a semi-arid Aegean island in Greece. In the first section, the importance of water resource governance and participation is reinstated. This is followed by a detailed presentation of water resource issues and conflicts in the case study island of Naxos. The suitability of scenario workshop as a participatory planning/evaluation tool is justified and a short review of related applications is presented. The second section presents in detail the methodology of a scenario workshop as applied in the case-study. This includes a presentation of both preparatory activities as well as the structure and processes of the actual workshop event. The outcomes of the workshop are presented in the third section followed by a discussion of results in the fourth section. In the concluding section, the key findings from the application of the tool in relation to the goals of the ADVISOR project are summarised.

8.1. INTRODUCTION

8.1.1. Water Resources Problems and Governance

The *critical* status of water resources at a global scale was emphasized in 1992 at the Dublin International Conference on Water and the Environment. Experts and representatives from national, international, intergovernmental and non-governmental organizations concluded that '*The problems highlighted (in this conference) are not speculative in nature; nor are they likely to affect our planet only in the distant future. They are here and they affect humanity now*'(ICWE, 1992). Fourteen years later and two years after the World Summit on Sustainable Development (2002), current modes and rates of development, combined with a variety of factors such as pollution, agricultural intensification and trends towards more frequent extremes in weather conditions are altering the quality, quantity and geographical distribution of surface and underground waters (UN, 2003). All these factors have spatially varying impacts on economic development, social structure and cohesion, while in numerous countries they have spawned poverty, environmental degradation and in some cases even disasters.

The underlying causes of water crises have been attributed to bad or ineffective water governance (GWP, 2000). Water *governance* refers to a range of political, social, economic and administrative systems that regulate the

 Integrated Evaluation for Sustainable River Basin Governance. Edited by Nuno Videira, Giorgos Kallis, Paula Antunes and Rui Santos. ISBN: 1843391481. Published by IWA Publishing, London, UK.

development and management of water resources and the provision of water services at different levels of society (UNDP, 2002). *Good* water governance is achieved when the relevant bodies responsible for water have established an effective policy and legal framework to allocate and manage water resources in ways responsive to national, social and economic needs, while ensuring the long-term sustainability of the resource base and the preservation of aquatic ecosystems and ecological processes (GWP, 2000).

At an EU level the need for a more integrated water policy and legislation has been recognized by the European Council already since 1988. The need to incorporate the novel aspects of hydro-biological quality, quantity and water use opened in 1995, and following a lengthy period of revisions of existing water legislation and the elaboration of new proposals, the Water Framework Directive (WFD) (2000/60/EC) came into force in December 2000.

8.1.2. Participatory Processes and Sustainability

Traditionally, the decisions concerning water related plans and projects have been the responsibility of government bodies and expert scientists. The concept that environmental issues are best handled with the participation of all concerned citizens and that the sustainability of water resources requires participatory processes during the planning and the implementation phase of water projects (as well as ecosystem-based management) (UN, 1992, UN, 2003) has been widely accepted during the last decade. In fact, insufficient participation of the public in the process of policy-making is considered as one of the reasons why environmental policies are often ineffective (ECFESD, 2000).

Participatory processes in decision-making have been identified as basic elements of good water governance that enhance the quality and the implementation of decisions. Furthermore, they raise public awareness on environmental issues; hence, a better informed public has increased opportunities to express its concerns and a better chance to exert some form of control over public authorities in matters relevant to the application of environmental law.

The appreciation that sustainability requires participatory approaches is reflected in the evolution of European Union's environmental legislation since the mid 1980's. The first steps towards openness in matters concerning the environment were taken in 1985, with the Directive on Environmental Impact Assessment (85/337/EEC) which introduced the consideration of the public's opinion before the initiation of projects bearing a probable impact on the environment. A decisive step towards a more participatory democracy in environmental matters was taken with the Aarhus Convention (1998) on "Access to information, public participation in decision-making and access to justice in environmental matters" (signed by the countries of the Economic Commission for Europe UN-ECE) and its recent adoption in European Union legislation with Directive 2003/35/EC on "Public participation in respect of the drawing up of certain plans and programs relating to the environment''. The WFD incorporates the provisions of the Aarhus convention mandating public information, written consultation of all interested parties before the final decisions are taken, and active involvement in the elaboration, review and updating of the Management Plans.

8.1.3. Water resource issues in Naxos

The island of Naxos (18.000 inhabitants) is located at a distance of 110 nautical miles southeast of Athens. It is the largest island in the Cyclades complex, with a surface area of 428 km^2, a coastline of 148km and a maximum altitude of 1009 m (Figure 8.1. map of Naxos). The climate is semi-arid and the annual precipitation is low (396 mm/yr) even for Greek standards. Winters are mild and in summer the local wind currents combined with high temperatures result in very high evaporation rates. The rocky mass of "mountain Zas" literally cuts the island in two, from north to south, with 70% of the relief consisting of medium to high slopes and the remaining 30% being productive cultivated lowland mainly at the west and NW side.

Naxos was an important centre of Cycladic civilization in antiquity. Unlike the rest of Greece, it remained prosperous for most of the 13-19th centuries A.D. enjoying considerable autonomy under Venetian and Ottoman rule and exporting mainly olive, vine and mineral products, most notably emery (polishing stone) in the Mediterranean region. Several monuments in sacred ancient Greek locations (Delos, Delphi) have been constructed from Naxos' marbles, which are still extracted today. In the 20th century war-related recession led to migration of a significant part of its population to Athens and abroad. In the 1990s, Naxos followed the prevailing regional trend of mass sea tourism which prompted the dense urbanization of its western coastal zone (Figure 8.1). However, unlike the other islands, which depend mainly on tourism, Naxos has traditionally preserved an equally thriving agricultural sector.

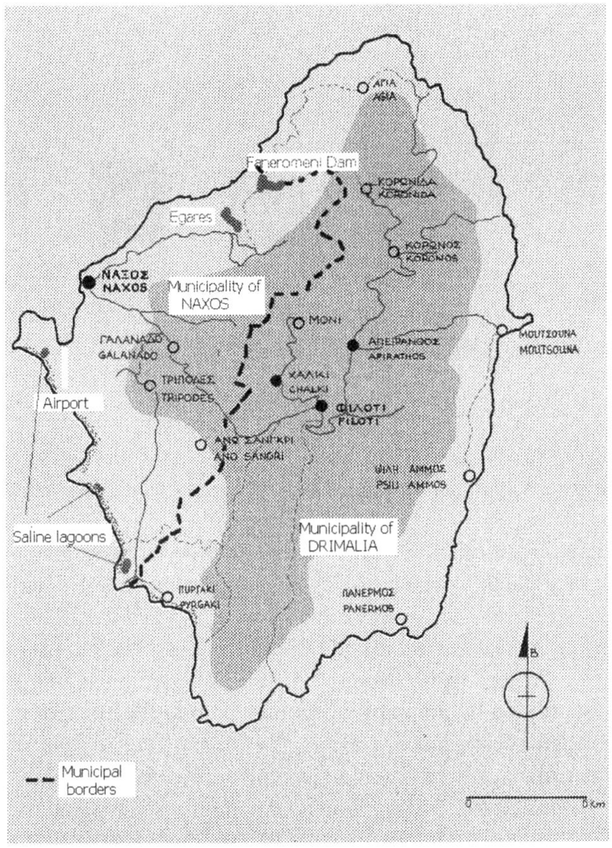

Figure 8.1 Map of Naxos and its two municipalities

Administratively, the island is divided in two municipalities, each comprising of smaller towns and villages. The municipality of Naxos (Figure 8.1), covering the western ¼ part of the island's surface area is the most prosperous, with a flourishing tourist sector (which occupies almost 50% of the workforce), the bulk of the agricultural exploitations (34% of the work force) and a relatively low, 7% unemployment rate. On the other hand, the mountainous municipality of Drimalia (including the villages of Chalki, Koronos, Moutsouna) is predominantly rural. The main economic activities include livestock breeding and small enterprises for milk and meat production. Drimalia has no tourist facilities. The unemployment rate (16%) is way above the regional and national average (6% and 10% respectively) (Daskalakis et al. 2001).

Figure 8.2 shows population trends in the island post-1970. The increase in the municipality of Naxos especially in the 1990s when the tourism economy growth started is evident. In addition to the permanent population, some thirty thousand tourists and visitors reside in the island in the summer period, most in the municipality of Naxos. The population in Drimalia is relatively stable, albeit ageing. Notably, Drimalia has the most populated rural villages in the Cyclades complex; locals are very proud of their traditional ways of living and have a strong feeling of connection to their place.

Water resource conflicts relate to conflicts over economic development opportunities between the different economic sectors, and between Naxos and Drimalia. The differences in economic development between the two municipalities, a relative attraction, especially of the youngsters of the rural hinterland towards the tourist lowlands where there are more employment opportunities (Daskalakis et al. 2001) and the competition for water resources by agriculture and tourism have generated tensions between the two municipalities and between farmers and the tourist sector.

Overall, water use in Naxos in 1997 was estimated at around 13.3 hm^3, (0.8 hm^3 for stock-raising and 1.5 hm^3 for drinking water supply and the remaining to agriculture (Daskalakis et al. 2001). Twenty five percent of the cultivated land is irrigated, a percentage three times the Cyclades' average demonstrating the continuous importance of the primary sector. The most important agricultural products grown are vegetables and fruit, with a productivity that covers local and occasionally national market needs (e.g. for potatoes and potato seed). Water needs for the very important livestock raising sector (76.0000 goats and sheep) represents only 6% of total abstractions. The island's meat and dairy products represent 38% of the total productivity in the Cyclades (Daskalakis et al. 2001).

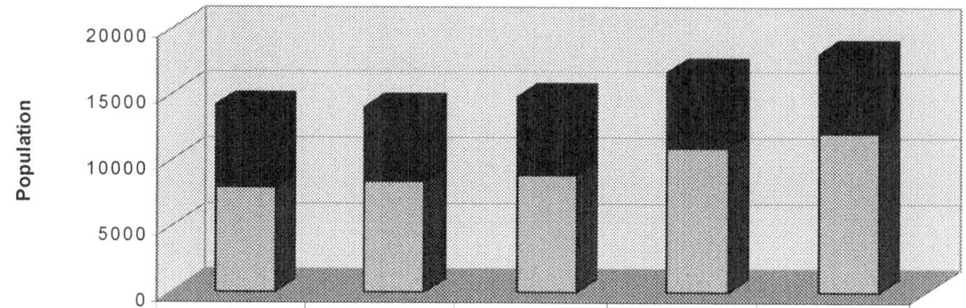

	1971	1981	1991	1997	2001
Municipality of Drimalia	6259	5622	5914	5756	6099
Municipality of Naxos	7942	8415	8924	11000	12089

Figure 8.2 Population changes 1971-2001 (data from National Censuses)

Gross estimations and models suggest that overall water availability surpass demand (estimated total water resources = 60 hm^3/yr). The majority of available water however is in underground deposits; Naxos has rich groundwater resources and numerous springs. Overall availability figures mask the fact that the karstic formations (mainly marbles) which contain these underground reservoirs are located in remote and isolated areas of the central and eastern parts (e.g. Moutsouna, Figure 8.1), at great distance from the agricultural exploitations and tourist developments of the west coast (Figure 8.1). Drilling and transportation by pumping of groundwater to the areas of greatest demand, or to the upstream mountain villages are very expensive (Hatzilacou, 2003).

Government efforts for water management have focused on surface waterworks. Two reservoirs were financed and constructed by the Ministry of Agriculture for irrigation (but ending-up serving mainly municipal uses) in recent years. The reservoir at the location of "Egares" (storage volume 600.000 m^3, Figure 8.1) was built in 1992. It is within the boundaries of the municipality of Naxos, which has the responsibility for regulating and allocating water among the various users (farmers including).

The second water work is an - almost completed - dam at the location of "Faneromeni" (storage volume 1.500.000m^3, Figure 8.1). There are plans for an irrigation network to distribute the water from the dam to farmers, but these have not progressed. The dam is at the borders of the two municipalities meaning that they both have rights over its management.

In summer, when demands for irrigation and drinking water reach simultaneous peaks, (Figure 8.3) partly because of the two-fold population increase linked to the seasonal tourist influx the existing reservoir is unable to cover demands.

Water shortages are common. Conflicts between farmers and the municipality of Naxos for the allocation of water abound. In the summer of 2000, supply of water to farmers from the two existing reservoirs was reduced to secure availability for coastal settlements and tourism facilities; this generated intense tensions. One year later, persistent drought prompted the central government to order an emergency transportation of water supplies to Naxos with the use of tankers.

The prospects of water shortage have decreased in the last two years due to the increase in rainfall (indeed, on February 2003 the island was stuck by torrential rains which caused massive destruction to farmers, livestock raisers and road networks), but are expected to resurface once the dry years return.

Most villages of Drimalia depend on local streams and wells of a limited quantity, highly dependant on seasonal variations. In the dry year of 2000 several mountain villages were cut off from central water supply due to shortages for 15-45 days. However, some villages of Drymalia are claiming the future exploitation of the newly constructed reservoir for municipal water supply and irrigation in the NW, an area which might be designated as mild tourist development zone and eventually will add to the income of its residents (Hatzilacou, 2003).

Allocation of water from the newly constructed dam at Faneromeni epitomizes these differences between the municipalities of Drimalia and Naxos and between farmers and the tourism sector. Although the cost of pumping water to mountainous villages should be excessively high, the municipality of Drimalia demands from the State to finance transfer works, or at least, for the municipality of Naxos to pay a compensation if Drimalia gives up "its" water rights from the reservoir. Furthermore, although the reservoir had been planned (at least on paper) by the Ministry of Agriculture for irrigation, the lack of an irrigation network to transfer water to the farms of the island is put as an excuse by the municipality of Naxos which plans to use the water exclusively for the drinking water supply of the coastal settlements.

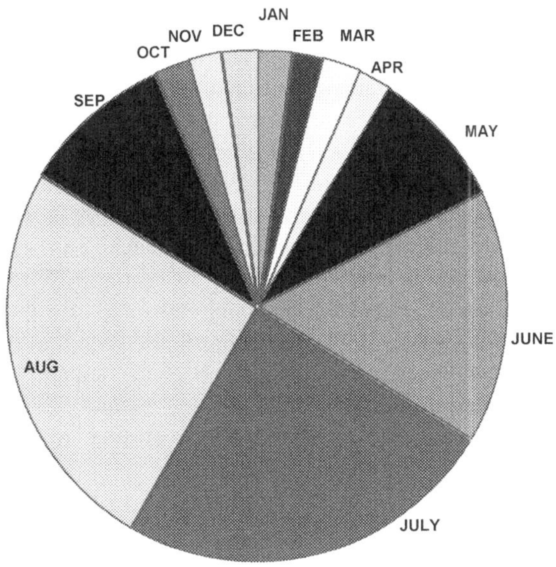

Figure 8.3 Distribution of annual water consumption in the Municipality of Naxos (municipal data)

Water shortages compromise the quality of local water resources. Several boreholes have been opened to pump coastal reservoirs, a practice which has led to over-exploitation and salinization of the underground aquifer, even 1,5 km inland. In 50% of the communities, sewage is disposed directly into small rivers or currents, thus polluting the nearby boreholes, wells and springs which supply drinking water to villages (Hatzilacou, 2003).

Water resource management in Greece is decided at central level as far as financing and construction of new water works is concerned, while network operation/maintenance, pricing and water allocation is the responsibility of the municipalities. Regional authorities are designated since 1987 by law as responsible for river basin management, but this power has not been materialised to date (the situation might change with implementation of the WFD). Regional and municipal administrative structures have limited financial resources, are ill-staffed and expertise is limited. Furthermore, the general lack of a national water strategy, or of any medium to long-term strategic planning at the regional level, is reflected in the case of Naxos island, where water management is a mere response to demands as they arise, on a "first come, first served" basis. Tourism, being the most dynamic and lucrative activity in the island, drives water management according to its growing demands, with other water users, mainly agriculture, lagging behind. The lack of a planning and allocating mechanism for the scarce water resources among competing use(r)s is stark, exacerbating conflicts and antagonisms.

8.1.4. A Scenario Workshop (SW) for Water Resources Management in Naxos

Given its specific characteristics, i.e. problems of water quantity and quality, potentially rich but costly to develop underground reserves, a diverse economy in transition and the emergence of conflicts among water users, Naxos was considered by the research team as a representative "miniature" case-study of water issues in Greece. Furthermore, Naxos had the additional feature that it is a small island (with limited water resources, flexibility in economic opportunities, etc), a situation poorly addressed by the Water Framework Directive, but very characteristic of the Greek reality of implementation.

The research objective was to use and test the tool of future "scenarios" as a deliberation platform for integrated, participatory evaluation and planning. A scenario is a picture of a probable situation or condition in the future. It describes alternative future options in a coherent, internally consistent way, while making assumptions regarding economy, advances in technology and innovation, social structure, and changes in mentality. When used in environmental planning, the basic aim of a scenario is to identify factors that will need to be taken into account in current policies in order to converge to desirable future scenarios and avoid undesirable paths.

The interest in scenarios had two origins. First, the long-term character of scenario-making was seen as a challenge in the context of Greece, where decision making tends to be ad hoc and short termist. The hypothesis was that by using a future oriented deliberation platform, extra benefits could be yielded in terms of initiating a longer-term dialogue in the community and supporting WFD planning procedures. Second, scenarios were seen as an ideal platform upon which to mingle and integrate water with development (economic, social, demographic) issues. A theoretical basis for the research was that water conflicts are primarily conflicts over development opportunities (Roberts and Emel, 1992) and hence water management should be framed as "part and parcel" of different development paths to the future. Scenarios provided a promising platform to examine water management and development issues in such a holistic way.

Scenarios have been used in conventional strategic and environmental planning based on experts or stakeholder groups (Mexa, 2002). More recently, scenarios have been used as an instrument to facilitate public dialogue (Rotmans et al. 2000; Guimarães Pereira and Funtowicz, 2003). "Visioning" is a tool which has seen applications in public and private organisations (Weisbord, 1993), urban planning (Walzer, 1996, Okubo, 1997), technology assessment (Andersen and Jaeger, 2001) and planning for urban sustainability (Street, 1997). Visioning workshops have been used as a tool for communities (in terms of representative stakeholders) to debate on their future and agree on a common of actions. "Future search" conferences are a methodological variant of visioning (Weisbord, 1993, Weisbord and Janoff, 2000). Future search has been applied on a variety of community cases and also for the preparation of a water quality plan in the Colorado river in U.S.A (Rehm et al. 1993). Visioning/future search events are based on a 2-day workshop structure. During the first day participants reach a shared vision about the "system" (company, community, etc) and in the second day, they elaborate this vision into a set of specific actions. Future Search differs from community visioning in terms of some specific tools used during the first day of the workshop in order to facilitate the resolution of a "common ground" between the participants.

A "Scenario Workshop" combines the use of scenarios with the process of a visioning (future search) workshop. The SW method originated on technological assessment (TA) studies (Mayer, 1997); visioning techniques and scenarios were conceived as proper tools in order to facilitate public debate and appraisal of new technologies (Street, 1997). In the early 1990s the Danish Board of Technology, extended the technique focusing particularly to environmental issues and the need to debate publicly new and integrated ways of handling environmental problems (Andersen & Jaeger 2002). Several SWs have been performed in European cities as part of an EU initiative on urban ecology and the role of technologies in the city of the future (see http://www.cordis.lu/easw).

A SW is a meeting in which a group of citizens discusses and interacts with other actors, exchanging knowledge and experience with the aim to develop a common vision and ultimately to produce a plan for future action with the widest possible acceptance. The focus is to create a framework for dialogue among policy-makers, experts and ordinary citizens while, in the process, new knowledge is created (Street 1997, Andersen & Jaeger 1999, 2002, www.panataneto.co.uk). The basis of SWs is a set of scenarios which are prepared beforehand and presented during the workshop to the participants, in order to stimulate their imagination, to generate new ideas and to assist the formulation of a common vision, the underlying concept being that a desirable future (an image of aspiration) which is shared by many, can be a powerful motivation for making this future come alive.

The standard application of SWs is a 2-day event with 28-32 people, invited and divided the first day in four groups: policy makers, scientists, experts and civil society. These homogenous groups work to present their own scenario and vision (goal) statement. The predefined scenarios are used as the platform for group work. Scenarios are framed upon two dimensions: responsibility for action (who: collective/public driven vs. market/individual) and tools/technologies used (how: low-tech/human-focussed ones vs. high-tech/technology focussed solutions). Four scenarios are given to participants corresponding to each of the four combinations of the two axes. Scenarios are accompanied by simple story narratives (e.g. a day in a household for each of the scenarios). With the use of facilitation, the four scenarios are debated first in groups and then in the assembly in order to reach a shared vision about the future. Like in a Future Search or a community visioning exercise, the second day is devoted into elaborating this vision into a concrete set of measures and in defining precisely responsibilities and times for implementation tasks (Street, 1997).

In the context of Naxos, a SW provided an appropriate and structured platform for running a scenario-based visioning exercise and hence testing its validity for initiating a long-term planning thinking on the management of water resources in the spirit of the WFD and for broadening the dialogue, between both the competing parties and the local society. A manual on the method (Bilderbeek and Andersen, 1994) was used to guide the structuring of the event. However, there were also a few concerns with the prototype Danish method and its applicability in the Naxos case. Modifications were introduced where necessary, first, by enriching the method with useful ideas from community visioning and future search and second, by adapting it to the specificity of the case (water) and place (Naxos). In the next section, the methodology applied is presented in more detail. The SW at Naxos is the first time a SW is applied on a topic relevant to freshwater management on a small island.

The specific objectives of the SW in Naxos as set at the beginning of the process and presented to participants are shown in Table 8.1. The basic aims of the SW held in Naxos was to assist participants in formulating a commonly accepted vision about how they wanted their island to be in the year 2020, with special reference to water and its sustainable management, and to specify ways or ideas (technological applications, policy directions, administrative and socio-economic arrangements, private initiatives a.o) in order to achieve the realization of this vision.

Table 8.1 Objectives of the Scenario Workshop in Naxos Island

(1) To gain new knowledge on issues pertaining to water and the environment
(2) To increase the insight and understanding of water related problems and issues by non specialist participants
(3) To achieve a dialogue among all participants and a sharing of views and ideas
(4) To help water specialists and decision/policy-makers gain a more integrated view of what lay people perceive as water problems and how they think about them
(5) To identify, discuss and achieve an appreciation of the differences and resemblances of problems and their solutions, as they are perceived by all participants
(6) To identify common solutions/ideas to bring the vision into reality and with the aim to resolve water related problems in Naxos
(7) To make available the results of the SW to the local authorities, regional media and to all the participants

8.2. METHODOLOGY

8.2.1. Preparatory Phase

Data gathering

The preparations for the SW in Naxos began one year before the planned workshop event (October 2002). The coordinating team identified the objectives of the workshop, the gaps in knowledge concerning water-related problems in the island, determined the kind of data that should be gathered and prepared a working plan. First, an internal working document was prepared, containing all the information gathered from an extensive literature review and collection of missing data about current and future development trends in the island, condition of water resources, environmental problems, etc (see: information summarised in section 8.1.3 above). The purpose of this document was three-fold: to bring the entire research team to a common understanding of the socio-economic context and water-related problems of Naxos, to assist the preparation of the four scenarios on "Naxos in the year 2020" which would be the central axis of the SW, so that they would relate with the island's reality, and to adapt the workshop's methodology to local needs.

A number of interviews were carried out in January 2003 with local stakeholders and scientists specialized on Naxos' water issues. These were pivotal in terms of understanding local water management practices and issues and acquainted the research team with the contrasting views of the various stakeholders. Furthermore, the interviews served to identify the network of the most important stakeholders that should be invited in the workshop. Interviewees were selected from the regional and local administration, the various economic sectors, citizens' organizations, NGOs and active citizens.

Interviews were based on a predetermined set of questions regarding water consumption in the main economic sectors, relevant conflicts, municipal water supply and sewage networks, sewage treatment, and others. Each interviewee was informed about the aims of the ADVISOR project and about the specific objectives of the SW. Upon completion of each session, the interviewee was asked to recommend other persons or bodies who should be consulted on these issues and to suggest additional relevant literature.

Preparation of scenarios

Four scenarios were prepared, based on the available information on the current situation of water resources in Naxos and on future development trends (Hatzilacou 2003). Each one described a radically different picture of Naxos for the year 2020. There were three key differences in scenario design from the original Danish version of the method (Bilderbeek and Andersen 1994):

1. The emphasis of Danish SWs on "technology" was dropped in favour of debating more generally the "issue" and the "system" (water resource management in Naxos), much alike the spirit of the Future Search method. Indeed, the application of SW on urban sustainability cases has demonstrated that the initial focus on technology was defied in practice by the participants who extended discussion and action plans to include several non-technological interventions (behavioural, social, institutional, regulatory, etc) (Street, 1997).

2. The second innovation of the methodology, which distinguishes it from SW (as well as Future Search applications), was that the scenarios (and in relation, the first day of the workshop) did not focus exclusively on the theme of water resources management per se, but on the wider development perspective for the island in the future, of which water management was a subset. This 'deviation' from the original method reflected the appreciation that any discussion concerning water management in Naxos inevitably would have to take into consideration development trends in the main economic sectors (agriculture, stock raising, tourism), as well as development perspectives in the rural uplands vs the lowlands. In other words, future water management schemes and approaches would depend, to a large

extent, on which development path the island would follow in the forthcoming future and vice versa, the choice of water management solutions would affect future development paths of the island. This however posed also some methodological challenges since there was the danger that it could be more difficult for participants to connect from first day discussions about Naxos in general to the proposal of specific water management measures in the second day.

3. A more multi-dimensional design of the scenarios was followed, in the spirit of the use of scenarios in planning (Mexa, 2002). The who/how dimensions of the prototype SW were kept but only as two among several other economic, social, environmental dimensions and descriptors used to frame future scenarios. Each scenario evolved along five dimensions: a) external forces (e.g. modernization of the administration, European Union expansion, global climate change, etc) b) trends in sectors of economic development (tourism, agriculture, mining, etc) c) population and spatial distribution, d) water resources and e) natural environment (Table 8.2). The water resources dimension was specified in terms of technologies and organisation, but the divisions of the Danish model (e.g. on collective vs. market organisation) were found inappropriate for the Naxos' context (e.g. there are few opportunities for market initiatives in the water sector in Greek islands, and a market-focussed scenario may appear alien to the participants). Distinctions were made in terms of whether management of water resources in Naxos would rely mainly on large-scale works, smart technical solutions, water-saving techniques or radical behavioural changes that would reduce water consumption and whether water management will be under the responsibility of the central state, the region, the local municipalities or a combination of the aforementioned.

Each scenario was one and half page long, and conveyed positive feelings to the reader, describing the future outcomes for the society of Naxos from an optimistic perspective. Special attention was given in this as the authors were prone to have some pejorative expressions regarding the contents reflecting a personal opinion (e.g. pro-environmental or pro-development). Furthermore, while writing the scenarios confusion emerged over whether the scenarios were meant to be "predictions" about the future and the impacts of different paths or if they were meant to be normative "visions", albeit improbable perhaps to be realised in the future. This confusion between the positive and normative character of scenarios has been identified as an important methodological problem in scenario design (Mexa, 2002).

In the process of developing the scenarios it was agreed that the four scenarios should be: sufficiently distinctive; desirable, each from a different normative perspective; internally coherent; possible under certain conditions, even if improbable given current trends. Each scenario therefore represents a different "world-view" and hence development perspective, ranging from the technology-centred/market globalisation one (S2) to the radical ecology / direct democracy one (S4). Therefore the use of scenarios was primarily "normative" in the sense of elucidating and allowing debate over alternative desirable futures, but also "positive" in the sense that these futures were not unfounded but fully justified in terms of possible trends and changes in external factors.

Indicative data was used to support semi-scientifically each scenario. For example, a graph with different water demand forecasts was prepared for each scenario (differences resulting from different assumptions on economic, population and per capita water use developments) (Figure 8.4).

Four "imaginary" letters were prepared and attached to the scenarios with the objective to present the scenarios' contents in a more pleasant and personal way since the technical form of the scenarios could be more difficult to be apprehended by some SW participants. Each letter was written by a hypothetic visitor, who returned to Naxos after an absence of 20 years and described to a friend the observed changes and general impressions. Finally, the scenarios were also prepared in the form of a PowerPoint presentation containing only pictures for the workshop in order to have a direct impression on the participants.

Mobilization of the local society and selection of the participants

Motivation of community / participants and presence of powerful actors are important factors for the success of a participatory event. The initial visits to Naxos for the collection of information and the series of interviews with local bodies established a close contact with the local society. An international workshop part of the ADVISOR project was decided to be held in Naxos in order to inform and aware policy makers and the community and "spread the word" of the upcoming SW. The international workshop and the SW were publicized by the local and regional media.

Furthermore, special attention was put in convincing the invited participants on the seriousness of the venture and the commitment required on their behalf. Three weeks before the SW, all selected participants where contacted officially by telephone in order to confirm their attendance. They were informed that participation was obligatory for both days and that a daily fee of 30 Euros was foreseen as a reward for working time lost in participating in the workshop. Ten days before the event, a package containing an officially-written welcoming letter, the SW program, an informative brochure on the *Implementation of the WFD* and the four scenarios with the attached "letters" were sent to each participant as a preparatory step in order to familiarize them with the scenarios (the backbone of the entire procedure), the objectives and the program flow of the SW.

Table 8.2 Concise version of the four scenarios

The "**business-as-usual**" scenario **(S1)** was for growth of mass tourism, a shrinking of the primary sector coupled with increased water demand which is served by big water works. Freshwater management is under the responsibility of the Central government, while water is occasionally transported from Athens during water shortages. This takes place in a context of trend changes in the rate of growth of the Greek economy, the modernisation of the public administration, the process of European unification and the role of Greece in the EU.

According to the "**modern**" scenario **(S2)** there is economic modernization of the island, globalization-fuelled growth with the use of cutting edge water technologies. The primary sector is reduced but productivity increases with the use of modern farming techniques and crops. Tourism addresses higher international incomes and high level and luxurious facilities are provided. Water management belongs to the regional administration, while the use of technology (desalination, recycling and reuse, remote sensing) enable effective qualitative and quantitative monitoring and management of water resources. This takes place in a context of a fast modernising Greek economy and public administration, in intensely "globalising" world markets and rapid technological progress.

In the "**balanced**" scenario **(S3)** there is a balance between economic development, preservation of cultural heritage and environmental protection while the tourist sector attracts a variety of visitors (naturalists, scientists, archaeologists and conventional). The local administration is responsible for water management, while the emphasis is on water conservation and small-scale, appropriate technologies. This takes place in a context of stable economy and effective governance, with a growing emphasis on policies for sustainable development and environmental protection. Worldwide markets for "green products" (farming, eco-tourism etc) have grown considerably.

The "**self-sufficiency**" scenario **(S4)** is a radical ecology one. The economy is a communitarian agricultural one (resembling production patterns of the 1950s) and relies on self-organization and sufficiency at the household level. The primary sector covers only local needs, tourism has shrunk and as a result there is a dramatic reduction in water consumption. Water resources are managed at a household and community level. Communities organise on a direct democracy basis. Decisions on water are taken by all citizens in open assemblages. This scenario is positioned some time after a global economic crisis whereby EU and global markets collapsed fostering a return to local sufficiency models.

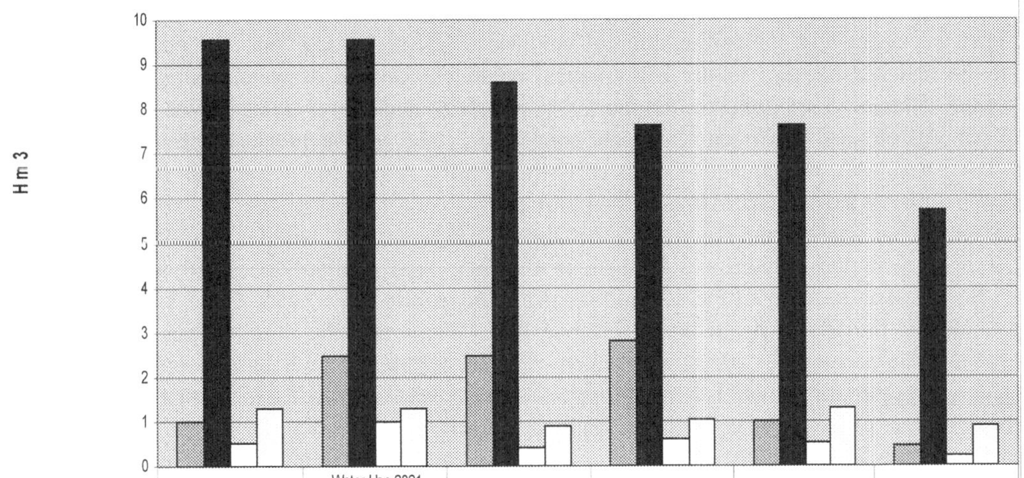

	Real water use (1997)	Water Use 2021 Forecast in Cyclades' Water Master Plan	Water use 2021 Scenario S1	Water use 2021 Scenario S2	Water use 2021 Scenario S3	Water use 2021 Scenario S4
Urban water supply M.Naxos	994238	2461470	2461470	2806076	1008000	450000
Irrigation M.Naxos	9555000	9555000	8599500	7644000	7644000	5733000
Urban water supply M.Drimalia	517685	994517	414148	607405	525628	225000
Irrigation M.Drimalia	1280000	1280000	896000	1024000	1280000	896000

Figure 8.4 Water Demand Forecasts 2021

Functionality of discussions and safeguarding the possibility of participants to express their opinions within the tight time limits were the two determining factors of group size. Whereas scenario workshops run with 28-32 participants (4 groups of 7 or 8), future searches and community visioning exercises have been ran with more participants (80-100). As this was the first time the research team was implementing the method, a bigger assembly with several smaller groups would be more difficult to handle. A maximum of 36-40 participants (i.e. four working groups the first SW day, with 9-10 persons per group) was decided.

Selection of participants was based on a "stakeholder analysis", i.e. recognition and invitation of important stakeholders by the organisers. The objective was to get the "whole system in the room" (Weisbord and Janoff, 2000), i.e. get in the workshop all important actors in the island with sufficient inter-dependencies and who will be needed if and when implementing a water resource plan. Drawing from the Danish SW approach, stakeholders invited fell into one of the following categories:

- Members of the public administration/ decision or policy-makers i.e. they have an active role in the design and implementation of local policies
- Scientists familiar with the local reality, covering the sectors of municipal water distribution or sewage or irrigation or hydrogeology of water management
- Representatives of the economic sectors (primary, mining, tourism and services),
- Citizens/ representatives of social, environmental or cultural NGOs, representatives from the Church.

While adequate representation on the basis of demographic, income or other criteria was not possible, care was given to have an as much as possible representative presence of women and younger people and a balanced representation from the two municipalities (Naxos, Drimalia). Due to size limitations the workshop was not open to "self-invited" participants. However, care was given to accommodate as much as possible in the citizens' group individuals active in community or water issues who approached the organisers before the SW and declared their interest to participate. Nonetheless, there were more availabilities than positions, and some offers had to be declined.

Organisation of the event

Practical issues, that might be underestimated from a "scientific" point of view, can be decisive factors for the success or failure of a participatory event. An internal *Manual* was produced for the members of the organising team describing all the details regarding the two-and-a-half day Workshop, e.g. the composition of the SW organising team, specific duties of each team member, a detailed program flow, written and visual material to be distributed/ presented to the participants at each phase of the SW, logistics. Its main objective was to ensure that all stages would be carried out according to the strict schedule of two 8-hour sessions. A "test event" for the entire procedure was also held by the organising team in Athens with the participation of volunteer students "acting" on imaginary roles. The research team prepared all the material relevant to the SW (papers, notebooks, magic drawers, pencils, tags etc) as these were specified in the *Manual*. Furthermore, diagrams, hydrological maps of Naxos, tables of water availability and abstractions per hydrological basin and other data concerning water management on the island were prepared in the form of posters. These were subsequently displayed in the working rooms and in the main plenary session room during the Workshop in order to assist the activities of the participants.

The selection of an appropriate facility to host the SW sessions was another important issue. Given that the procedure required open plenary discussions as well as work in groups, a High School with several sufficiently spacious rooms was selected. The location of the school was symbolic, located in Chalki, the old capital of the island at the geographic center of the island, in an equal distance from Naxos and the mountainous villages of Drimalia, and practical, enabling the access of participants from the various communities. A large room with the capacity to host simultaneously 40 people sitting in a U shape (for visual contact) was selected for the plenary sessions and four smaller ones for the working groups. Other factors considered during the selection of rooms were luminosity (pleasant environment), acoustics (avoidance of echoing during discussions), being on the same level (to avoid translocation of participants between floors during shifts from group to plenary sessions) and the availability of a large space outside the rooms where participants could gather and discuss between sessions.

8.2.2. Workshop Sessions

The SW for the sustainable management of Naxos' water resources took place in the public high-school of Chalki, Naxos on the afternoon of the 31st of October and the 1st and 2nd of November 2003. It was attended by 36 participants. Tables 3 and 4 present a concise outline of the two day SW. A 3-hrs working afternoon (Friday, 31st) where participants were introduced to the goals and processes of the workshop and were given the opportunity to express their aspirations and contribution preceded the 2-day main event.

Table 8.3 Agenda of the first day

1st Day Visioning	
Plenary session	
09.30 – 09.40	Objectives of the Scenario Workshop
09.40 – 09.55	Presentation of the day's program
09.55 – 10.00	Introduction to the notion of SWs
10.00 – 10.40	Visual presentation of the four scenarios
10.40 – 11.00	Guidance to working groups
Work in Groups	
11.00 – 13.00	Formulation of a "VISION of Naxos in the year 2020" by each of four groups (following a structured procedure)
Plenary session	
13.00 – 14.00	Presentation of each group's VISION
14.00 – 15.30	Discussion, debates, conclusion
Lunch break	
Plenary session	
16.40 – 17.30	Formulation of a common vision for Naxos in the year 2020

Creation of Common Vision for Naxos

The 1st day was dedicated to the creation of a common vision for Naxos in the year 2020 with special reference to water resources. The ultimate objective was to help participants assess and link future water needs with specific choices regarding socio-economic development and the environment of their island. The four groups were labelled with colours (red, blue, yellow and green). The participants were divided in to four homogenous groups but the criteria of assignment into role groups were adapted to the local context. To the criterion of role in the community (administration, economy, experts in water issues, citizens) which is used by the Danish method, the criteria of economic interest (farmers vs. tourism business) and place of origin (Naxos vs. Drimalia) were also used as distinguishing factors, given their importance for water management in the island. Representatives from the economic sector and policy-makers were therefore mixed in two groups according to a rural vs. coastal distinction. The *Red group* consisted of representatives or people employed in agriculture, stock raising and policy makers of the municipality of Drimalia (mountain uplands). In the *Blue* group, participants were representing the tourism sector and policy makers from the municipality of Naxos. The other two groups were like in the Danish method. The *Yellow* group, represented citizens' associations and NGOs. The *Green* group consisted of scientists and experts in the wider field of water resources management.

Each group was asked to prepare its own vision. A questionnaire (7 questions on future conditions in the island written in a simple and specific way) prepared by the coordinating team for this occasion could be used as guidance in group work. The list included 5 optional questions of general interest (addressing issues of population, employment, culture, environment etc) and two mandatory ones which addressed water quantity and quality issues (to ensure that the vision would address the issue of water). In comparison to the Danish method where participants are asked to work on the scenarios (read them, propose modifications, etc), a more flexible approach was followed in Naxos. Participants were given the option to draw ideas from the scenarios without being constrained on those. It was more in the group facilitator's discretion when, how and to which extent to refer to the scenarios.

Each group presented its vision (with a title / slogan) in the plenary and representatives from the four groups remained for an afternoon plenary session (after the break) to formulate a common vision providing a fairly balanced representation of all four groups. a factor deemed necessary in order to reach consensus over the final wording.

Table 8.4 Agenda of the second day

2nd Day Idea Generation	
Plenary session	
09.20- 09.40	Presentation and approval of COMMON VISION
09.40 – 10.20	Status of water resources in Naxos (expert presentations)
10.20 – 11.00	Modern trends in freshwater resources management - Examples of best practices (expert presentation)
11.00 – 11.15 :	Creation of four thematic groups
11.15 – 11.30 :	Break
Work in Groups	
12.40 – 13.30:	Idea generation – voting of 3 best ideas per group
Plenary session	
13.15 – 14.00	Presentation of 3 best ideas per group (in total: 12 ideas)
14.00 – 15.00	Discussion
15.00 – 15.30	Voting of five best ideas (among the 12)
Lunch break	
16.30	Evaluation – Closing ceremony

Idea generation

The second day was devoted to the generation of new ideas with the aim of formulating an action plan. Three specialists on water management and very familiar with Naxos' water issues were invited to give lectures, so that participants could all start from a minimum level of common information and knowledge. Participants were then divided into four thematic groups, each dealing with one of the four themes that emerged from the Common Vision regarding water resources (technological solutions/waterworks, water saving, measures for water quality, administrative arrangements). Groups were intentionally heterogenous containing a mélange of participants from all the groups of the 1st day (homogenous groups in the first day are intended to form stakeholder group visions, while in the second day synthesis and partnership is sought). Participants were encouraged to generate a list of ideas and preferences. A "Samoa Circle" was used whereby the floor passed to each group participant who should propose shortly in one sentence a measure. This facilitation method has the advantage of speed, getting out specific proposals as well as getting all participants active in the process. The initial recording of all proposals after several rounds was followed by more detailed discussion on each one, where those who proposed them were asked by other members of the team to clarify them. Connected measures were grouped together. A voting process then followed in group to choose those considered as most important the best three qualifying for a detailed presentation in the assembly. Then the assembly voted for the top 5 among the 12 selected measures.

After the voting sessions and the selection of the five best ideas (Top Five) participants were asked to evaluate the entire process by filling-in a circulated questionnaire. This evaluation served as essential feedback for the coordinating team.

In the prototype method as well as in future search and community visioning workshops, the identification of specific measures/ideas is followed by a last task in group where participants work voluntarily in the measure they wish, developing a time table of implementation, foreseeing tasks and resources needed, etc. The results of these groups are integrated in an overall Action Plan. Ideally, this last group work has the potential of building long-past partnerships around the implementation of specific measures/projects (Weisbord, 1993). This last session was dropped out in the Naxos SW for reasons explained in section 8.4.

The duration and structure of the Naxos SW had to be adapted to the actual capacity of the participants to attend. Following the request from several participants to reduce the daily duration of the SW, some tasks initially foreseen were cut in order to allow for the shorter timetable presented in Tables 3 and 4. For example a common-ground building exercise in the 1st morning (taken from Future Search), whereby participants would identify external factors of change for Naxos (hence cultivating a sense of "commonness") was dropped out. Group work was also shortened and less time was given for discussion and debate. Finally, an afternoon plenary session for the formulation of the common vision became optional, rather than mandatory for all, as was the initial intention.

8.3. WORKSHOP OUTCOMES

Participants were asked to express in writing their expectations from the SW during the welcome session the afternoon preceding the actual workshop. A concise list of the most frequently appearing expectations, which emerged from 8 working groups and their fulfilment (based on an evaluation procedure at the end of the SW) are shown in Table 8.5.

Table 8.5 Degree of fulfilment of participants' expectations from the Naxos SW

Expectations raised by participants	F.O. *	% of participants satisfied (based on questionnaires)
Gaining Knowledge	14	84
Experience on participatory processes	4	85
Truly open and free dialogue	6	84
Implementation of the outcomes	5**	-
Other	3	-

*F.O: frequency of occurrence of the specific expectation, **: to be shown in the longer term

The evaluation questionnaire which was distributed upon the closure of the SW, was answered by 25 of the initial 32 participants (seven participants had to depart earlier). The analysis shows that expectations were to a large degree satisfied, which can explain the sense of optimism, contentment and the positive comments during the closing ceremony. The largest success was the respect towards other participants' views and the achievement of open dialogue.

The Common Vision

The visions of the four working groups and the Common vision that emerged from the identification of the common elements among them are shown in Table 8.6. Overall, the four visions had several similarities but also differences;

however the most commonly emerging idea was a diversified economy, instead of the prioritisation of only one sector. The preservation of traditional occupations and activities, combined with appropriate technological solutions, protection of the natural environment and conservation of the rich cultural heritage was also a common aspiration. Concerning water management, small to medium scale water works and an emphasis on water saving in a variety of forms was a common denominator. The shared vision was described by some as a hybrid of scenario 3, with a touch of technology from scenario 2 and a desire for autonomy and self-sufficiency from scenario 4. This reflects a desire for developing without loosing tradition and for cooperation between different activities and communities.

Agreement on the Common Vision was a difficult task, especially since participants gave different meaning to expressions like "balanced" or "sustainable" development and were not willing to surrender ground on the exact definition put in the vision. Participants were reassured that all opinions (including those not included in the final wording) were recorded and would be included in workshop proceedings.

Table 8.6 Summary of group visions and Common Vision

Red Group-Man & environment with respect to development
Lively villages populated throughout the year. Mild economic activities and tourism. Protection of the environment and cultural heritage. Small dams, reservoirs, water saving. Institutional framework.
Green Group- Development within limits
Population increase, youth, employment in all the island. Modernization of production, labelled products. Land-use planning, protection of the environment and rehabilitation. Water works, locally based water management body.
Blue Group- Tradition and future
Young people, employment, populated uplands. Variable economic activities, modernized agriculture, tourism, commercial enterprises. Preservation of the environment. Water efficiency, quality protection, Local water management body.
Yellow Group –Quality of life
Life, variable economy. University. Technological applications in agriculture, mild development. Traditional occupations. Protection of natural environment and of cultural heritage. Water-saving, Local water management body.
Common Vision
We dream of a Naxos in the year 2020 with a diversified economic development, of medium scale and with quality, with improved modernized infrastructures. With educated and responsible citizens, who possess knowledge about sustainable development. A Naxos center of knowledge and education, with University institutions contributing to the life of the island, with more self- sufficiency, where there is dialogue and cooperation between citizens and institutional bodies. With harmonic co-existence between man and the environment, with protection of the most important and sensitive natural ecosystems. We dream of a Naxos with sufficient water quantity and quality, with the appropriate technologic but also traditional techniques, with demand management for water saving.

Idea generation

Eighty three ideas and proposals emerged from the thematic groups of the second day, others being very specific and detailed in nature, while others were more general. In addition to technological options, participants' proposals covered institutional (inter-municipal water management body, compensation fee, education), legal (allocation rules) and environmental matters (preservation of terraces, local laboratory for quality control). Furthermore, interesting and innovative ideas were put forward not only by the experts but by all the participants. The three best ideas per thematic group and the five best ideas that emerged form voting in the plenary session are shown in Table 8.7. Measures voted most demonstrate an emphasis on small scale waterworks (e.g. a new dam or a desalination plant were not voted favourably, whereas protection of terraces, small rural reservoirs and water saving measures were on the Top five), monitoring of water quality (a very poorly performed task, the importance of which is underplayed by authorities) and to "soft" measures in terms of allocation rules and education programmes.

Instead of the last SW session on the preparation of an implementation plan, an alternative concluding session was performed whereby participants were encouraged to express their ideas for a follow-up to the SW. The idea was to seek for individuals or representatives from participants who would be willing to undertake the responsibility to develop some of the measures shortlisted and develop a partnership for its implementation. Proposals included, among others the allocation of funds from both municipalities for the continuation of the SW procedure, the organization –by local bodies- of meetings for the educating the public on specific issues and the creation of a local research laboratory, staffed with specialized scientific personnel and supported by volunteer citizens and other bodies.

Table 8.7 Best ideas per thematic group (number of votes per group, and in plenary session)

Water Saving	**Water quality and Environment**
• Education, public awareness, mentality change on water saving (13) • Metering, Licenses for ground water pumping (11) • Subsidized water-saving devices for homes-water saving in hotels (7)	• Sewage treatment in all Naxos (7) • Pesticide & nutrients: awareness and reduced application (7) • Water quality control and creation of a laboratory for all the island (6)
Administrative measures	**Technological solutions**
• Distinguish in planning and allocation between drinking water for local purposes and drinking water for tourism (7) • Locally based water management through cooperation between the two municipalities (5) • Compensation fee from Naxos to Drimalia communities for the water it uses from the Faneromeni dam (5)	• 3-4 small upland reservoirs for Drimalia (9) • Improvement and extension of municipal supply and irrigation networks (8) • Protection and preservation of earthen terraces to improve water infiltration (7)
Top five (plenary session)	
• Protection and preservation of earthen terrasses to improve water infiltration **(21)** • Distinguish in planning and allocation between drinking water for local purposes and drinking water for tourism **(20)** • 3-4 small upland reservoirs for Drimalia **(20)** • Water quality control and creation of a laboratory for all the island **(19)** • Education, public awareness, mentality change on water saving **(19)**	

8.4. DISCUSSION

In terms of the expectations expressed by participants (Table 8.5) which overlapped -more or less- with the seven objectives laid down during the first day, almost all were satisfied according to an analysis of responses to the evaluation questionnaire. 78% of respondents felt they gained new knowledge (objectives 1, 2) and 84 % expressed deep satisfaction with the quality of the dialogue (obj. 3). Participants spoke openly about their concerns and most of them felt they had the opportunity to appreciate problems from different perspectives. This is confirmed by the acknowledgment by 96% of the participants that dimensions previously neglected emerged from the workshop process (obj.4). Furthermore, 84% felt that, after the workshop they had more faith in the use of dialogue and in participatory processes as a means to solve important problems; they also had more understanding and respect towards different opinions (obj. 5).

Results of the questionnaire however have to be treated with caution. Participants may be prone to respond positively due to a "feel good" factor after the event or they might feel that they have to be polite to the organisers. Furthermore, the success of the event cannot only be appraised in terms of whether participants (or organisers) were satisfied with it (Shipley, 2002). Much more is expected from participatory processes as part of the WFD, not least delivery of practical outcomes and implementation changes. Scrutiny and theoretical reflection on processes and results of the workshop follows below.

8.4.1. Implementation outcomes

The Naxos SW fell short of producing a real Action Plan for water resource management or generating specific and long-lasting implementation partnerships for some of the measures. A main reason for this was that the workshop was run as a one-off research experiment and was not part of a broader policy initiative.

As Wood and Gray (1991) note high stakes are a necessary condition for successful collaboration, and hence participation in a collaborative process. Indeed in the Naxos SW there was great willingness on the part of the public to participate and serious commitment of important policy actors (though not all) in the event. Several factors can explain this strong positive response: shared concerns (torrential floods inflicted severe damages to the entire island nine months earlier, combined with a series of years of drought and water shortages), curiosity about the process, desire to "be heard" (by the people for politicians, and by the politicians for citizens) or perhaps, the professionalism and the meticulous preparation of the workshop by the organisers (including development of personal relations with invited participants), etc.

Despite this interest, the Naxos SW was not integrated in a real policy or decision-making process. The promise from the organizers to participants that the results from the workshop will be "forwarded" to regional and national authorities tried to account for this, but it was insufficient. Furthermore, the process was not initiated from the local community (a crucial factor of success; see Weisbord, 1993), but from the organizing team who wanted to test the method. Therefore one can hypothesize (and there was evidence in the meeting to support this hypothesis) that participants may have felt that their opinions and the vision or measures selected would not really matter, hence putting less energy in the process. This can have two effects. Less powerful participants may become frustrated and retreat from a process that they see as a continuation of the "status-quo". Powerful participants may be prone to show a "good face" and agree to radical ideas, knowing that the agreement will have little real impact. This might partly explain the agreement on a rather radical vision in the Naxos case, far from actual developments in the island. The representative of hotel owners initially reacted to the vision because it did not recognize the need for development of conventional tourism facilities on the coast (emphasis was on cultural, agro and eco tourism in the rural areas). He did not feel however that he needed to "fight" to get his ideas through the final text. This might give the impression of a false consensus in the vision phase, which hides however a lack of real willingness to implement agreement.

Recognizing the distortion of the lack of a real policy context for the Naxos' experiment, still there was room for testing some of the methodological elements of the SW and to discuss whether they were conducive or counterproductive for achieving agreement over actual implementation. Shipley (1997) in a critical ex-post evaluation of visioning exercises argues that the lack of successful implementation follow-ups relates to foundational weaknesses of the method (such as the general character of vision statements). A careful reading of the experience from other SWs reveals that few, if any, really arrived at an Action Plan which was then implemented (see for example, Street, 1997). Methodological weaknesses of the SW which may have limited the achievement of an Action Plan in Naxos too include:

- There are no established mechanisms for conflict resolution and consensus (e.g. voting, etc) over the agreement on the common vision. Indeed, the SW methodology focused mainly on consensus building, a fact which may have given some stakeholders a veto power over specific visionary reform (Heiling 1998) thus blocking the emergence of more concrete statements and objectives. This means that real choice seldom takes place, and there is a tendency to include all proposals in the vision, hence leading to a rather general statement. The watered-down and general wording of the common vision, putting forward general strategies and goals rather than concrete objectives and measures can be seen as a drawback in terms of implementation. On the other hand, the future development of an island is a general issue to begin with. Added to this, the questions which participants relied upon to "build" their vision were also of a general nature. Had the questions been more precise in order to "lead" to more specific choices, it is possible that more concrete objectives would have emerged, at least from the group visions. In other words, the specific tools used during the visioning exercise (e.g. choice of questions) in the 1st day can determine to a certain degree the final outcome.
- Difficulties in comparing and voting between ideas/measures. Some of the measures proposed in the Naxos SW were very specific in nature, while others were more general, or contained a subset of several more ideas. Participants were asked to choose between partly incomparable entities.
- There is limited time for the necessary clarification, elaboration in depth and deliberation over the numerous ideas and measures that are proposed, both in the thematic groups and the plenary sessions. In fact, lack of sufficient time was highlighted by participants in Naxos as the workshop's most serious disadvantage. Achieving in 2 ½ days what was not achieved in years may be over-ambitious.
- 87% of participants in Naxos SW felt they needed more technical and scientific input in order to chose among ideas they were unfamiliar with (see more on this below) in order to make an informed choice and vote. Lack of time and the feeling by participants that they did not have enough information led to the decision by the organisers to skip the final session, where an implementation plan for each of the Top five measures would be worked on separately.
- Proper representation and democratic legitimacy. There were complaints by some participants that the composition of the workshop was not "representative" of the population of the island and hence the vote had no real democratic value (more on this below).

On the other hand, some potential positive practical outcomes of a SW emerge, if seen in context:

- Even if general, still the vision highlighted a certain development path and declined others. For example, the Naxos SW vision stressed diversity and "balanced" (however defined) development over "mono-cultural" tourism development of the coast. In a context where the vision would be used for real-life decisions, this could be a useful guidance. For example, a "balanced development" vision would rule out at least some extreme water allocation solutions (such as the current practice of prioritizing tourist facilities without questioning their demand). Nonetheless a problem is that apparent consensus in common vision may simply mask the still underlying differences that will surface only on implementation.

- The voted solutions may be general, but still useful if they are reframed and seen as "ideas to be taken up" by authorities or partnerships "for further consideration" (that was the way the task was reframed during the SW after complaints from participants against calling the High 5 solutions an "agreed action plan"). The results of a SW should not necessarily be seen as a direct-democratic end-output for an Action Plan, but as a first step of goal setting (i.e. vision) and measures' identification in a process of consultation and deliberative decision-making. The ideas for example in Naxos' SW for restoring land terraces to retard and allow infiltration of rainwater or for public education programs on water saving could be taken up by the authorities and studied together with conventional solutions (waterworks, etc). Then a proposed Action Plan by the authorities could be returned to a community forum for further elaboration and evaluation (e.g. by a social multi-criteria evaluation process or other deliberative method).
- Agreement on a comprehensive Action Plan should be distinguished from the goal of a formation of one or more partnerships around specific implementation tasks. For example, in comparison to the SW, the Future Search method is designed more towards identifying and implementing "win-win" actions, where partnerships between stakeholder participants can be built (Weisbord et al. 1993). Note however, that agreement on actions of common interest (e.g. an all-Naxos school education program for water saving) does not mean that conflicting issues have been resolved (e.g. allocation of water from the Faneromeni dam).

8.4.2. Information and the Role of Scientific Expertise

As mentioned above, many participants complained that they did not have enough information or knowledge on the issues in order to make a proper judgment. Water resource issues are complex and require a degree of technical and scientific knowledge that is not available by ordinary citizens. Although in many participatory processes citizens feel they need more input from "those who know", giving to scientific expertise a privileged status in the participatory process can disempower citizens, intensify feelings of ignorance and lead them to defer to perceived authority (Holmes and Scoones, 2000). For example, in the Naxos SW the decision to add presentations by "experts" on the morning of the second day to provide participants with some basic information proved counterproductive. The partial positioning of the experts in the local controversy about the allocation of water from the new dam generated reactions and nearly destroyed the co-operative atmosphere of the first day.

An interesting research topic is the development of information enhancing tools to support a SW process. For example, a separate information session with alternative views presented could have preceded the SW and/or an accompanying analytical report written and given to partners. Note however, that in the case of Naxos (which is also characteristic of many other places in Europe and beyond), data available was limited, unreliable and uncertain. There are a few analyses and only a handful of "experts", all engineers. Gaps in knowledge and uncertainty in data are an inherent characteristic of water resource studies, especially in countries such as Greece with poor data monitoring systems.

Quality assurance methods (see Chapter 5) could also be potentially integrated with SWs. The very nature of an SW however, focusing on a broad vision, may render difficult (and to a certain degree irrelevant) the detailed treatment of scientific information, complexity and uncertainty.

Requirements in terms of information will depend on the actual goals of the SW. If the goal is to formulate an Action Plan, then lack of information is a serious barrier. If however, the aim is to raise awareness / motivate the participants, enhance dialogue, set goals and preliminary ideas or ideally or activate some "win-win" implementation partnerships, then lack of detailed and reliable information may not be such a decisive obstacle.

8.4.3. Representation

The nature of stakeholder and citizen representation is closely linked to the accountability and legitimacy of the entire process (Holmes and Scoones, 2000). There are four basic options in selecting participants for a participatory process/event: identification and invitation of selected stakeholders (i.e. those who matter); open invitation / "self-selection" of participants; systematic selection upon predefined representative criteria (e.g. gender, age, income group, profession, etc); random selection from the population.

Selection of participants clearly depends on the purposes of the exercise and the design of the participatory event. For example, adequate representation may not be that important if the goal is education. It will be crucial however if the goal is to reach a consensual decision. Inclusion of powerful, marginalised or reacting voices will be important if the goal is to reduce conflict. Certain methods can work with an open number of participants but others face constraints.

In the Naxos SW a pragmatic approach based on stakeholder invitation was used with some allowance for self-selection and an effort in the selection list so that the group was not very unrepresentative of the characteristics of the population (especially in terms of gender and age). Stakeholder representation was a good choice for the goal of "bringing the whole system in the room" in order to find "win-win" partnerships (Weisbord, 1993). It proved however less appropriate for reaching an agreement on an Action Plan. Indeed, some participants in the Naxos SW reacted explicitly during vision-making and in the voting process because they felt that the result lacked democratic legitimacy. Hence the organisers reformulated the name given to the task to "ideas to be further looked upon" rather than "agreed actions".

Another problem with stakeholder representation is that it tends to lead to the use of "known and established networks which may not represent the broader community of users" (Sanderson 1999). For example, some sectors of society, perhaps systematically excluded from such processes in Naxos, as for example elder people, handicapped citizens, Albanian immigrants or the poor were not invited in the SW because they did not appear as important "stakeholders" in water resource issues. If the goal of the process was to get marginalised voices in, then a very different selection procedure would have to be followed.

There is no standard recipe or methodological blueprint for the participants' selection process. Different rules and approaches apply to different goals, methods and contexts. Organisers of a SW should be aware and explicit of the limitations of any selection procedure.

8.4.4. Facilitation and Framing

The fact that group facilitators did not interfere during the elaboration of the common vision (apart from registering emerging thoughts and assisting their proper grouping) and by "religiously" remaining impartial to all expressed views may have enhanced participants' trust in the dialogue process and motivated their involvement. It was during the plenary sessions, where the SW facilitator's role became more decisive and evident: resolving conflicts and extracting the underlying common denominators, diverting the dialogue towards more productive paths and adjusting the speed of the process (slowing down upon the need for further elaboration, accelerating upon repetitions) were some of the most important factors that enhanced the quality of the dialogue. This, on the other hand reflects the facilitator's "power" to influence the dialogue outcomes, by framing problems, or by further elaborating certain issues according to his(her) theoretical and scientific background and his(her) sensitivity to specific issues (e.g. environmental or "sustainability" related ones).

Organisers are also often "actors" (external or internal) in the policy process, with their own perceptions, interests and agendas. Their role in framing issues and defining objectives of the process is at odds with the idea of an unbiased debate. For example, the very decision on whether there is an issue meriting a participatory process and the very act of choosing long-term visioning as the way to frame responses are important framing choices. It is very different to frame a community debate generally on water policies (as was done in Naxos) and very different to focus it on the determination of ecological objectives for the water bodies (as is required by the WFD). The latter already presupposes that the need for ecological objectives is accepted; this however, may not be always shared by the local community.

This hints to an apparent tension between the commitment of the WFD to specified objectives on the one hand, and participation, on the other. A participatory exercise such as a SW rests on the idea that participants are free to debate and determine their own vision and goals. The WFD on the other hand sets very specific, ecology-related goals which might not be shared by communities, certainly not in socio-cultural contexts such as those of Naxos (where there is little environmental awareness and limited freedom in terms of economic opportunities). Constraining public discussion on how to achieve the pre-defined goals of the WFD may run counterproductive to the very idea of engagement and participation in the process.

8.4.5. Dialogue and Deliberation

In his quest to discover truth, Socrates (500 B.C) set the foundations of the dialectic method, a process which could be accomplished only through cooperation with others. After pretending to accept the validity of a reply to his question, by applying dialogue he ingeniously guided his co-speakers to unfounded conclusions, thus forcing them to admit to their ignorance. In other words, he developed what Bloomfield et al. (1998) call "a reflective capacity that enables the evaluation, re-evaluation and eventually the transformation of initial values and preferences". This is the essence of "deliberation" through dialogue.

Shipley (2002) questions the instrumental value of long-term visions and whether they contribute to the quality of decisions. Heiling (1998) also notes that setting process rather than outcome objectives may explain why some visioning exercises are not successful in promoting immediate action or planning. Both these analyses however have a different perspective than the one adopted in the present research which perceived deliberation and participation as containing value over and above the quality of the decisions that emerge (Holmes and Scoones, 2000). Outcome goals are no panacea; process goals might also be very important and they are a crucial element of the planning process. In this respect, the goal in this research was not only to check the usefulness of scenarios and visioning in terms of the effectiveness in decision-making, but also in terms of their contribution to promoting dialogue and deliberation per se.

The SW methodology managed well in facilitating dialogue between participants, especially in the elaboration of the common vision in groups and in the assembly. Participants found the scenarios relevant (95%), useful (100%) and constructive (90%) in the debate. The contents of the scenarios (accompanied by their vivid narration in present tense) liberated participants' imagination on the one hand, while helped them visualize better the link between water needs and the choice for socio-economic development, thus blending in the issue of "carrying capacity". Focusing on the carrying capacity of the island in terms of water resources, instead of dealing with the specific claims of interested groups has the advantage of encouraging shared understanding, thus avoiding conflict and leading to a more integrated approach to water management. For example, in the discussions many participants criticized those who referred to insurmountable barriers, claiming that much can change in twenty years if there is will. Also a high-level discussion took place on issues such as the meaning of "sustainable development", "governance", "participation", etc. This "maturing" of the dialogue between simple citizens should be seen as positive evidence of the role played by the SW.

The contribution of the SW in deliberation is evidenced in that 84% of participants felt they had more understanding for each others' opinions after the workshop. On the other hand, whereas a SW facilitates dialogue, the structure of the method is partly designed so as to avoid conflict, directing work to creative paths. This has the potential of agreement on win-win solutions, but is less capable of resolving deep rooted conflicts. For example, in Naxos, agreement in the need for an education programme was relatively easy but agreement on something that would affect the controversial allocation of water from the Faneromeni reservoir would be much more difficult. Further research is needed on how visioning and scenario workshops can be complemented or supported by conflict resolution techniques.

8.4.6. Time and resources

Time is a major constraint for an effective process. A 2 ½ days weekend event is not particularly long; in addition, compensation was foreseen for participants in the Naxos SW. Still many individuals encountered problems in attending all sessions. For example farmers had to irrigate their fields, hotel owners could not be away from their business for the whole weekend, mothers had to attend their children because husband was working on the field, etc. The programme was modified in situ and shortened to adjust to requirements of the participants. On the other hand, lack of time was recognised in the evaluation by participants as a major drawback. Improving the participatory process (e.g. by including an information workshop, etc) would need even more time. This highlights a key tension between the needs of participatory democracy and present socio-economic conditions.

Costs for the SW in Naxos included:

- Human resources: two post-doc level scientists for a total of 15 person/months + 2 person/months of secretarial support. One professional facilitator for the workshop (1 person/month) plus three volunteers/assistant facilitators for group sessions and one recorder and time-keeper (note that the workshop can also run with only 1-2 facilitators and self-organisation of group work).
- 2000 Euros in travel and subsistence for the organising team.
- 1000 Euros in consumables and material for the workshop.
- Compensation fees (30 Euros per day) for the 36 participants.
- 2000 Euros for two lunches (offered at discounted price by the local women's association).

For a one-off SW the amount of personnel work devoted in the Naxos experiment sufficed. Personnel costs will increase if a more ambitious and longer-term participatory process is sought (e.g. having more than one workshop, building constant relationships with community, follow-up of implementation, etc). Still, the budget appears relatively low and can easily be accommodated as part of a broader planning process or water management project.

8.5. CONCLUSIONS

The research objective of the SW at Naxos was to test the usefulness of the tool for integrated, participatory evaluation and planning as part of the implementation tasks for the WFD. An integrated evaluation process takes into account the various dimensions of the issues, involves all different perspectives and participants that express it, and assures the quality of information used.

Does a SW deliver? A response is "deliver exactly what"? Shipley (2002) has a point when he notes that visioning exercises confuse different goals such as awareness, learning, participation, long-term thinking, innovative ideas, consensus and action, that cannot be achieved all in once. Indeed, the Naxos SW experiment revealed that as applied, SW worked well as a platform to foster awareness, mutual learning and dialogue and opening-up the public debate. The method also performed well at mapping key issues and alternatives and getting bottom-up input, i.e. from some people who are usually excluded from such debates (e.g. teachers, students, individual citizens). There was multiplicity of ideas and proposals emerging beyond the conventional technocratic solutions, highlighting a broadening of the decision space. On the other hand, SW had considerable methodological weaknesses if sought to deliver an implementation Action Plan or resolve conflicts. In relation, an important finding is that different goals demand different choices of SW design (in terms of choice of participants, framing of issue, provision of information).

In terms of the first dimension of integrated evaluation, i.e. information, SW seems ill equipped in treating information complexity and uncertainty. This however should not be a big problem if the goal of the process is not agreement over specific measures/Action Plan. Information support tools can be used to enhance the scientific quality of the SW.

In terms of assessment, the SW performed well at mapping key issues, enabled the inventory of ideas and the drafting of a first list of alternative solutions. However, in order to take the process on step further, i.e. to prioritize and select ideas for further action, the SW method is limited. Potentially a SW could be used (e.g. with a devoted last session) in a deliberative identification of the necessary assessment criteria (technical feasibility, economic viability, environmental sustainability, etc) for a selection procedure "further down the process". Constraints in terms of time, representation of participants, knowledge and information do not make SW an appropriate tool for the actual multi-dimensional assessment of alternatives.

Finally, as far as participation and deliberation are concerned, the SW method facilitates and creates an enabling environment for structured dialogue between participants and it permits the expression of views of otherwise marginalized actors and groups. On the other hand, the quality of the dialogue achieved is deeply dependant on the qualities and personality of the SW facilitator, which in fact may, under certain circumstances become the Achilles heel of the process.

With relation to the specific implementation tasks of the WFD, the SW can provide a valuable tool for the early stages of the planning process. It can be used for example as a platform upon which to debate the "significant water management issues" in the river basin or to formulate goals in a river basin water resource (quantity) plan. The SW does not appear a promising method for the evaluation of derogations, specific projects or for the final formulation of the program of measures (although it can contribute to a "bottom-up" listing of potential measures). An apparent contradiction with the WFD is that whereas a SW rests on unbound vision-making, the WFD sets specific (ecological) objectives. Of course a SW could be designed in terms of different scenarios for the achievement of ecological objectives in the river basin in a long-term horizon (a different framing than the one in Naxos). Although this could be a useful tool say for a group of river basin stakeholders entitled with the implementation of the WFD, it is questionable whether broader communities who are unaware of the Directive could connect to such a narrowed goal, especially in local contexts such as those of Naxos.

REFERENCES

Andersen, I-E. and Jaeger, B. (1999) Scenario Workshops and Consensus Conferences: Towards More Democratic Decision-Making . *Science and Public Policy*. **26**(5), 331-340.

Andersen, I-E. and Jaeger, B. (2001) Scenario Workshops and Urban Planning in Denmark. *PLA Notes* **40**, 53-56

Andersen, I-E. and Jaeger, B. (2002) Danish Participatory Models. The Pantaneto Forum. Issue 6. April 2002. (www.panataneto.co.uk)

Bloomfield, D., Collins, K.l., Fry, C., and Munton, R. (1998) Deliberative and Inclusionary Processes: Their Contributions to Environmental Governance. Paper presented at the first ESRC Deliberative Inclusionary Processes in environmental decision – making seminar.

Commission of the European Communities (CEC) (2000) Proposal for a Directive of the European Parliament and of the Council on Public Access to Environmental Information., 29.06.2000 COM(2000) 402 final 2000/0169 (COD), Brussels.

Daskalakis, K., Aravosis, K., Toppos, Y., Bezes, K. (2001) Study on the water management of the Cyclades. Prefecture of the Cyclades. Technical Services.

International Conference on Water and the Environment. (1992) Stockholm Environment Institute Publications, Stockholm.

European Consultative Forum on the Environment and Sustainable Development (ECFESD) (2000) Sustainable governance: Institutional and procedural aspects of sustainability. European Commission.

Guimarães Pereira Â, Funtowicz S. (2003) "Methods for citizen involvement in new governance: reflections based on three empirical cases", *Technikfolgenabschätzung – Theorie und Praxis* **2,** 57-62.

GWP (2000), Towards Water Security: A Framework for Action. Global Water Partnership. Stockholm

Hatzilacou, D. (2003) Issues of water availability and management in the island of Naxos. Unpublished internal report . 16 pp. Advisor Project. University of the Aegean, Mytilene.

Heiling, A. (1998) Collaborative visioning: Proceed with caution!: Results from evaluating Atlanta's Vision 2020 project. *Journal of the American Planning Association.* Chicago: 1998. Vol. 64, Iss. 3; pg. 335, 15 pgs

Holmes, T. and Scoones, I. (2000),Participatory environmental policy process: Experiences from Norht and South. Institute for Development Studies (IDS) Working Paper 113.

Kallis, G., Videira, N., Antunes, P., Guimarães Pereira, Â., Spash, C.L., Coccossis, H., Corral Quintana, S., del Moral, L., Hatzilacou, D., Lobo, G., Mexa, A., Paneque, P., Pedregal, B., and R. Santos (in Press). Integrated participatory assessment for water resource planning and governance

Kallis, G. and Butler, D. (2001) The EU water framework directive: measures and implications, Water Policy, **3**,125-142.

Kallis, G, Hatzilacou, D., Svoronou, E., Coccossis, H., Xenarios, S., Tziritis, H., Mexa, A., and Symvoulidou, M. (2004), (in Greek). Scenario Worskhop for sustainable management of water resources: the case of Naxos. University of the Aegean.ISBN 960-87812-1-3. 135, Mytilene.

Mayer, I. (1997) *Debating Technologies: A Methodological Contribution to the Design and Evaluation of Participatory Policy Analysis* University Press, Tilburg.

Mexa, A. (2002) The Scenario Method in Strategic Environmental Planning (in Greek), *Topos,* **18-19**, 215-227

Okubo, D. (1997) *The Community Visioning and Strategic Planning Handbook,* The National Civic League, Denver, CO.

Regional Environmental Center for Central and Eastern Europe (1998) *Doors to Democracy- Current Trends and Practices in Public Participation in Environmental Decision-making in Central and Eastern Europe.* ISBN 963 8454 56 3.445 p. on-line version only (www.rec.org/REC/Publications/Publications_PublicParticipation.html

Rehm, B., Schweitz, R., Granata, E. (1993), "Water quality in the upper Colorado river basin", in Weidbord M (ed) *Discovering common ground* (Berrett-Koehler Publishers, San Francisco)

Roberts, R S., Emel, J. (1992) Uneven development and the tragedy of the commons: competing images for nature-society analysis *Economic Geography* **68** (3), 249-271

Rotmans, J., Asselt van, M., Anastasi, C., Greeuw, S., Mellors, J., Peters, S., Rothman, D., Rijkns, N. (2000) Visions for a sustainable Europe *Futures* **32,** 809-831

Sanderson, L. (1999) Participation and democratic renewal: from 'instrumental' to 'communicative' rationality? *Policy and Politics,* Vol **27** No3, 325-341

Selman, P. and Parker, J. (1997) Citizenship, Civicness and soclia capital in Local Agenda 21 *Local Environment,* Vol **2**, 171-184

Shipley, R. (1997) Visioning in planning: is the practice based on sound theory *Environment and Planning A* **34,** 7-22

Street, P. (1997) Scenario workshops: a participatory approach to sustainable urban living? *Futures* **29** (2), 139-158

United Nations Conference on Environment And Development (1992) *Agenda 2,* UN publications.

United Nations (2003) World Water Development Report –Water for people, water for life. Berghahn Books / UNESCO Publishing. 576p.

Walzer, W. (ed) (1996) *Community Strategic Visioning Programmes,* Praeger, Westport.

Weisbord, M and 35 International Coauthors (1993) *Discovering common ground. How future search conferences bring people together to achieve breakthrough innovation, empowerment, shared vision, and collaborative action* Berrett-Koehler Publishers, San Francisco.

Weisbord, M., Janoff, S. (2000) *Future search. An action guide to finding common ground in organizations and communities* Berrett-Koehler Publishers, San Francisco.

Wood, D.J., and Gray, B. (1991) Toward a Comprehensive Theory of Collaboration. *Journal of Applied Behavioral Science* **27** (2), 139-162.

Chapter 9

Mediated Modelling To Support Public and Stakeholder Participation in Water Resources Planning and Management: the Baixo Guadiana Experience, Portugal

Nuno Videira, Paula Antunes, Rui Santos, Daniel Borrego and Gonçalo Lobo

ECOMAN, Ecological Economics and Environmental Management Centre, DCEA- FCT, New University of Lisbon, Portugal

This chapter describes the application of a participatory modelling methodology to a case study conducted in the lower part of the Guadiana River Basin, referred to as "Baixo Guadiana". Mediated modelling was used as a tool to support the active participation of the interested parties in the scoping of the Baixo Guadiana problems, pressures and impacts. A stakeholder group collaborated throughout three workshops in the interactive development of qualitative and quantitative models addressing the main problematiques of this river basin. The paper aims to contribute with further insight on how to promote effective participation of the interested parties in water resources planning and management.

9.1 INTRODUCTION

The rationale for public participation in environmental decisions has been concisely expressed in Principle 10 of the Rio Declaration on Environment and Development: "environmental issues are best handled with participation of all concerned citizens, at the relevant level... each individual shall have appropriate access to information concerning the environment that is held by public authorities and the opportunity to participate in decision-making processes..." (UNCED, 1992).

Over the past two decades, public and stakeholder participation in environmental decision-making has been increasingly defended for enhancing: social responsibility, social learning, procedural fairness, the integration of social values into analytical decisions, public trust and confidence in decisions and decision-makers, and the quality of technical assessment processes (Petts, 2003). In alignment with the participatory principle, several Community Directives and international agreements have been proposed for strengthening the inclusion of public and stakeholder concerns in decisions related with environmental matters (e.g. the Convention on Access to Information, Public Participation in Decision-making and Access to Justice in Environmental Matters - United Nations Economic Commission for Europe and the Directive 2003/35/EC - providing for public participation in respect of the drawing up of certain plans and programmes relating to the environment).

 Integrated Evaluation for Sustainable River Basin Governance. Edited by Nuno Videira, Giorgos Kallis, Paula Antunes and Rui Santos. ISBN: 1843391481. Published by IWA Publishing, London, UK.

With legal provisions in place, the challenge for public and stakeholder participation is passed to the side of (effective) implementation. More so when empirical evidence shows that good practice goes far beyond granting a right to participate. Stated differently, improvements in practice lag improvements in law.

In the ADVISOR project, Videira et al. (2003a) have critically reviewed and compared the participatory elements of the evaluation processes of five water related projects in Europe. The main lessons from this research indicated that interested parties were involved in late stages of the evaluation process, there existed low levels of participation impact (mainly information and consultation instead of co-design), low evidence of influence on decisions, low deliberative impact of the participatory methods used (mainly public hearings and public comments) and that no concern from the project promoters was demonstrated in a formal assessment of participatory outcomes with respect to the ultimate goals of strong democracy.

Participation in water resource planning and management is crucial for good water governance. As highlighted by a recent study from the United Nations World Water Assessment Program (WWAP, 2003), the water governance crisis is caused by the lack of adequate water institutions, fragmented institutional structures, upstream and downstream conflicting interests regarding rights and access to water, diversion of public resources for private gain and the unpredictability in the application of laws, regulations and licensing practices, which affect markets. In a context of high complexity and uncertainty, participation is regarded as one of the basic principles for effective governance. Accordingly, clear and transparent rules should be set, supported by deliberative decision-making structures, which allow for a shared responsibility in safeguarding water resources.

In Europe, the future of water policy is being shaped by the Water Framework Directive (WFD) (Directive 2000/60/EC). Aiming at the protection and improvement of all European waters, the WFD introduces an innovative approach to ecological and integrated water status assessment, river basin planning, elimination of pollution by dangerous substances, implementation of financial instruments and public information and consultation. Lanz and Scheuer (2001) defend that the success of the implementation of the WFD will strongly depend, among other factors, on political will and on the full participation of all stakeholders. These authors claim that it is the integrative powers of the technical and political networks to be established which would determine the success of the WFD.

The main provisions for public participation under the WFD are centred on Article 14. According to this article, the public (including users) has the right to be informed before and during the planning process, the right to comment and the right to have access to the background documentation and information on the future River Basin Management Plans (CEC, 2000).

Beyond information and consultation, stakeholders and the general public are encouraged to be "actively involved" throughout the whole WFD implementation cycle. The potential benefits for active public participation in the context of the WFD implementation include: the possibility to obtain a better identification and consensus on the key water management issues; the integration of knowledge, experience and concerns of local communities in the RBMP; an increase in commitment toward the Programs of Measures; the potential to reduce implementation costs of the WFD; and improved enforceability of both regulatory and voluntary approaches (if developed in partnerships with stakeholders) (WWF, 2001).

Although its potential benefits are undisputed, is not clear how active involvement of the interested parties will be promoted. During a recent DG ENV/EPRO Seminar[17] on the implementation of the WFD, a number of uncertainties were raised: what is "active involvement" and what can authorities do to encourage it?; at which scale(s) should public participation be organized: local, regional, national, river basin level?; how to reach the unorganized public?; how to promote social learning?; when providing relevant information to interested people, who frames the issue?; public motivation should be ensured before the issue becomes critical; the use of innovative techniques should be explored.

As part of the WFD Common Implementation Strategy (WFD-CIS), it was recognized the need for developing practical guidelines for the implementation of the different aspects of the Directive. Several working groups were established to develop non-legally binding and practical guidance documents, addressing different technical issues of the Directive, which are targeted to the experts who are directly or indirectly involved in its implementation. The informal Working Group on Public Participation, established under the Working Group 2.9 (on Best Practice for River Basin Planning) has published in December 2002 the "Guidance on Public Participation in relation to the Water Framework Directive – Active involvement, consultation and public access to information" (ComEC, 2002a). This document is to be tested and validated in pilot river basins selected across several EU Member States, including 4 countries of the ADVISOR project consortium: Greece – Pinios River Basin; Portugal – Guadiana River Basin

[17] Proceedings of the Seminar: "Implementing the Water Framework Directive: Implications for Regional Authorities. 4th June 2003, at the Steiermark Regional Office Jointly organised by DG Environment and the Environment Platform of Regional Offices (EPRO). Available at the Circa Website http://forum.europa.eu.int/Public/irc/env/wfd/library

(Portuguese side); Spain – Júcar River Basin; The Netherlands (along with Belgium and France) – Scheldt transboundary River Basin.

This chapter describes the participatory modelling case study that was conducted in the lower part of the Guadiana River Basin, known as "Baixo Guadiana". Participatory modelling was used as a tool to support the active participation of the interested parties in the scoping of the Baixo Guadiana problems, pressures and impacts. It aimed to contribute with further insight on how to promote effective participation of the public and stakeholders in water resources planning and management. The case study was developed as part of the second phase of the ADVISOR project that focused on the development and testing of tools to support integrated planning and evaluation of river basin interventions, in the context of the WFD.

In the following section the fundamental principles of the participatory modelling methodology are presented. The third section describes the design of the participatory process implemented in the Baixo Guadiana, whereas section 9.4 focuses on the discussion of the results achieved in each of the steps. The final section summarizes the key findings from the application of this tool.

9.2 MEDIATED MODELLING

Mediated Modelling (MM) is a form of participatory modelling which aims for a collaborative team learning experience to raise the shared level of understanding in a stakeholder group, while fostering a broad level of consensus. MM allows for overcoming the shortcomings inherent to linear thinking and nonparticipatory decision-making, when tacking complex dynamic problems (van den Belt, 2004).

MM draws on the principles of the system dynamics methodology (Forrester, 1961) - a problem oriented method in which a given system is defined in terms of the underlying structural relationships. In system dynamics, the fundamental assumption is that the behaviour of a problem arises from its structure. Sterman (2000) defines the structure as "the feedback loops, stocks and flows, nonlinearities and delays created by the interaction of the physical and institutional structure of the system with the decision-making processes of the agents acting within it" (p.107).

In the majority of cases, user-friendly software packages (e.g. Dynamo, Stella, Vensim or Powersim) assist the modeller in building the stock-and-flow structure of a system dynamics simulation model. These programs also facilitate the modelling process in the formulation of its mathematical equations. With drag-and-drop actions the different types of variables are selected from the menus onto the working window, and afterwards clicking on the variables will prompt dialogues for setting up the equations and parameter values. At the end of this process, the model may be run to observe behaviour over a given time horizon. The building blocks of structure include stocks (which represent accumulations), inflow and outflow rates (into and out of a stock) and auxiliary variables (which help defining the inflow and outflow rates). The computer program then performs the numerical calculations (i.e. simulation) in order to generate dynamic behaviour over the specified time horizon.

System Dynamics has been used for direct modelling with client groups since the 1980's (Vennix et al. 1992), as an approach to facilitate team learning, decision-making and policy change (Zagonel, 2002). This approach was called "group model-building" and has been widely used in corporate organizations Vennix (1996). Van den Belt et al. (1996) proposed MM as a similar approach where participation is broadened to representatives from several stakeholder groups affecting or affected by environmental problems.

The most common "tangible" goal of MM is to construct a computer-based simulation model at the "scoping level", i.e. a high-generality and low-resolution model that serves to increase the understanding of the dynamic behaviour patterns at play rather than predicting a precise outcome. Eventually, the scoping models may be further iterated to more detailed and complex research or management models.

Thus, a MM process consists of a series of modelling workshops where a broad stakeholder group collaborates in the development of a simulation model, which is aimed at scoping and exploring innovative solutions to complex and dynamic problems. A mediated modeller, using visually oriented modelling techniques and software, facilitates the workshops. Apart from the modelling workshops – the core platform for deliberation with this tool – MM involves preparatory (identifying and setting-up the stakeholder group, conducting introductory interviews and preparing a preliminary model) and follow-up activities (interviews or questionnaires, reports and model tutorial).

Two types of models may be developed with system dynamics: qualitative and quantitative. Qualitative models consist of conceptual models depicting the main variables (and associated relationships) that describe a dynamic problem. These models may not be simulated and are only useful for problem definition and conceptualisation purposes. Quantitative models are achieved after the quantification of the model variables and require the synthesis of the knowledge/information disclosed by the participants into a computer-based model. These models are then used for policy analysis (typically during the final workshop) by comparing and evaluating the results of different "simulation runs", which correspond to alternative "what-if?" scenarios.

The unique features of the system dynamics methodology make it specially suited for a participatory exercise. The group modelling process allows for structured deliberation since the stages of the model building method (Richardson and Pugh, 1981) are integrated in a broader participatory process of reflection on the river basin issues. A logic and pre-defined sequence of tasks shapes the backbone of the agenda for the modelling workshops. The model building process encourages participants to focus on the leverage parameters affecting decisions (Stave, 2002).

This process is flexible in application to a variety of subjects and participatory contexts, although the tangible output is pre-determined, i.e., building a dynamic simulation model. Flexibility is also possible in the procedure itself. The modelling team may prepare different scripts for knowledge elicitation tasks that the facilitator could employ at critical stages, or distribute workbooks to participants providing detailed methodological explanations and preliminary results;

The initially complex jargon, i.e. the building blocks of the modelling method, is soon transformed in the common and open communication vehicle for all participants. It shifts attention away from rigid points of view and allows stakeholders to present their knowledge on the subject in a constructive and open manner. As argued by Stave (2002), such a collaborative modelling process levels the playing field during the evaluation of policy scenarios. It does so by allowing all participants to propose alternatives and then focus on the analysis of its consequences over time and across different model sectors. This promotes openness and facilitates collaborative policy design.

A systems approach is helpful for creating a holistic view in a team, by integrating piecemeal viewpoints. Participatory modelling promotes team learning and knowledge since the modelling process aids in storing, structuring and transferring the vast amount of ideas, knowledge and opinions ventilated in a meeting (Vennix, 1996).

The following section describes the design of the MM process implemented in the Baixo Guadiana river basin.

9.3. PARTICIPATORY MODELLING IN THE BAIXO GUADIANA

9.3.1. The Baixo Guadiana

The Guadiana river basin (total drainage area of 66800 km^2) is shared between Spain and Portugal (17 per cent). This river basin is one of the most studied in Portugal. In ADVISOR it had already been selected for the ex-post review of the Alqueva project evaluation process (Figure 9.1).

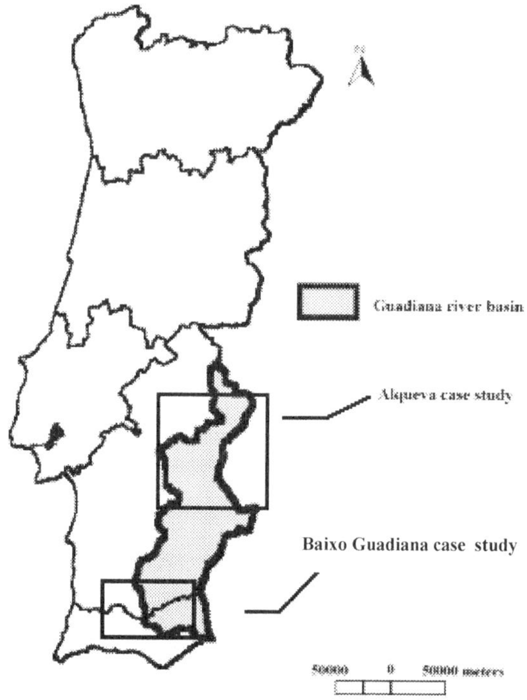

Figure 9.1 Location of the Baixo Guadiana case study area

The large number of dams built upstream, both in the Portuguese and Spanish sides, affects the water quantities arriving at the southern part of the basin. In the Baixo Guadiana, water plays a strategic role in the development of several activities such as domestic water supply to half of the Algarve region, fisheries, agriculture, salt making industry and tourism. The Baixo Guadiana also has a high nature conservation value. Located at the lower end of the Guadiana's estuary, the Natural Reserve of the Saltmarsh of Castro Marim and Vila Real de St. ° António was the first area to obtain such classification in Portugal (2089 ha designated in 1975). Three overlapping conservation areas have been established in this salt marsh: a Special Protection Zone under the Birds Directive and a Special Conservation Area under the Habitats Directive (both part of the Natura 2000 Network) and a Humid Zone under the Ramsar Convention. While relatively small, these protected areas have an important influence on the mosaic of economic activities.

9.3.2. Participatory modelling process design

Research objectives

The specific goals of the participatory modelling exercise included:

- The definition and implementation of a participatory methodology for scoping the problems, pressures and impacts in the Baixo Guadiana;

  To promote synergies with the process of implementation of the WFD CIS, the research team has met with the National Water Institute (INAG) - the ADVISOR Portuguese end-user. INAG is also the body from the Portuguese Ministry of Cities, Land-use planning and Environment which is responsible for the development and implementation of the national water policy. INAG has supported the selection of the Guadiana River Basin for the ADVISOR case study, provided information on the Pilot River Basin (PRB) testing process, and highlighted the potential contribution of the case study to the early "Characterization and Analysis" step of the implementation of the WFD, although the experience was not formally included as part of the activities developed by INAG for the PRB testing.

  The characterization step of the WFD implementation includes the review of the environmental impact of human activity in each river basin. The CIS "Guidance for the Analysis of Pressures and Impacts in accordance with the Water Framework Directive" (ComEC, 2002b) concluded that the public and stakeholders should be involved, as early as possible in the scoping stages of a planning and evaluation process, where different perceptions and values can still be incorporated in the identification of problems, goal setting and design of alternatives. Furthermore, "(…) the list of pressures and assessment of impacts on a water body…shall ensure the identification of all of the potentially important problems. Assessing the likely impacts arising from each of the pressures will produce a list…that then becomes a basis for developing a program of measures which might be undertaken in order to achieve good status….a screening approach is likely to simplify the tasks…and Member States should aim to achieve the best estimate of significant pressures in the time available…this could involve a combined approach of assessing monitoring data, model usage and expert judgment" (ComEC, 2002b). It should be stressed that although the process was aimed at the identification of the pressures and impacts in a participatory way, it did not explicitly focus on the assessment of the impacts in relation to the WFD objectives (including those for protected areas).

- The use of a Quality Assurance Protocol (QAP) for the assessment of the research activities planned;

  Following the results and recommendations from ADVISOR's Horizontal Report for WP2 on "Quality of Information in River Basin Governance – Review of case studies: Information Vertex" (Guimarães Pereira, 2003), one of the concerns of the research team was to account for the quality of information and procedures used in the participatory modelling experience. To this extent, it was developed a specific Quality Assurance Protocol (QAP) (Corral Quintana, 2000; Corral Quintana and Guimarães Pereira, 2004) for the assessment of the research activities planned for the participatory exercise implemented in the Baixo Guadiana case study.

- The development of a "common view", in the form of a causal diagram (Qualitative Model), of the current problems, pressures and impacts perceived by the participant group;

  This was possible due to the definition of a novel approach, which combined the typical computer-based MM process with the causal loop diagramming techniques foreseen in the group model-building method. Such common view was expected to be achieved at the end of the first deliberative event, integrating the participants' perspectives on the Baixo Guadiana river basin current problems, pressures and impacts. Another novel element which was tested consisted of linking the development of the causal loop diagram by a "Driving forces-Pressures-State-Impact-Response" (DPSIR) indicator matrix;

  - The development of a quantitative model in POWERSIM, addressing the problems, pressures and impacts identified by the group. The simulation of the model was expected to support the analysis of alternative development scenarios (e.g. management measures, interventions/projects);
  - The development (with the collaboration of the participant group, and as a consequence of the participatory modelling results) of a draft strategic plan, including strategic objectives and measures for the Baixo Guadiana.

Stages of the participatory modelling process

In a participatory modelling process the main interactive model-building stages include: problem definition, conceptualisation, specification, and policy analysis. Each consecutive stage is usually developed in distinct workshops, although there may exist several variants due to the flexibility of this tool. The length of each workshop also depends on a number of factors (*e.g.* number of participants involved, desired model detail, problem complexity, etc) although it typically varies from half a day to two full-day meetings. The optimum number of participants for a MM process is about 20-30, although these limits are merely indicative. Past MM projects have been conducted successfully with groups of 5-10 participants (in small-group sessions), while others registered more than 40-70 participants (in plenary sessions) (Videira et al. 2003b). The model building stages should be complemented by several tasks that are performed behind the scenes of group modelling. These tasks included the synthesis of intermediate results in workbooks to be distributed to participants at the beginning of each workshop reporting on the results and explaining methodologies used.

The process planned for the Baixo Guadiana included the following stages (see Figure 9.2 for an overview of the participatory modelling process):

- Preparatory activities;
- 1st Workshop (13 February 2004) - problem definition and development of the qualitative model;
- 2nd Workshop (29 March 2004) - conceptualisation of the quantitative model;
- Specification of the model was performed by the research team in-between the second and third workshops, using the data sources suggested by participants. It was also possible to organize small group meetings with experts for the specification of some model sectors;
- 3rd Workshop (2 June 2004) - analysis of alternative scenarios;
- Follow-up activities.

The research team has performed the typical roles foreseen for a MM process, including:

- The facilitator - responsible for eliciting knowledge and conducting the group discussions;
- The modeller - who develops the model in the computer;
- The process coach - reports to the facilitator on strategies to keep the group effective;
- The recorder - writes down and sketches the important parts of the group proceedings.

In MM it is also needed a person within the responsible/sponsoring organization for direct contact throughout the process. This person serves as the process gatekeeper, i.e. someone who usually helps to initiate the process and assists in framing the management problems and identifying the appropriate participants. The Director of the natural reserve performed this role. The natural reserve was also the venue for the MM workshops.

9.4. RESULTS AND DISCUSSION

9.4.1. Stakeholder identification and invitation

A group composed with members of the research team, the gatekeeper, and managers of the Natural Reserve was established for the organization of the modelling workshops. This steering group met regularly throughout the Baixo Guadiana case study. In the first two meetings the goal was to discuss the project objectives and generate an initial list of stakeholders to invite according to the following methodology:

- Identification of key issues in the Baixo Guadiana area;
- Generation of a key list of interested parties (institutions or groups but not names of representatives/individuals) associated with the key issues at stake;
- Identification of representatives from the identified interested and the corresponding address/contact information;
- Checking and validation of the preliminary list.

Invitations were sent to all the stakeholders included in the preliminary list, together with a leaflet explaining the project objectives. The invitation included a form to suggest other people they found relevant to participate, given the stated objectives. This cross-referencing procedure was used as a way to considerer possibly excluded groups and to validate the stakeholder list.

A set of mailing invitations was sent to organizations' representatives and other local actors (the majority of these are landowners inside the Natural Reserve) and those stakeholders then suggested others. The consolidated list of stakeholders invited throughout the project included a total of 86 persons.

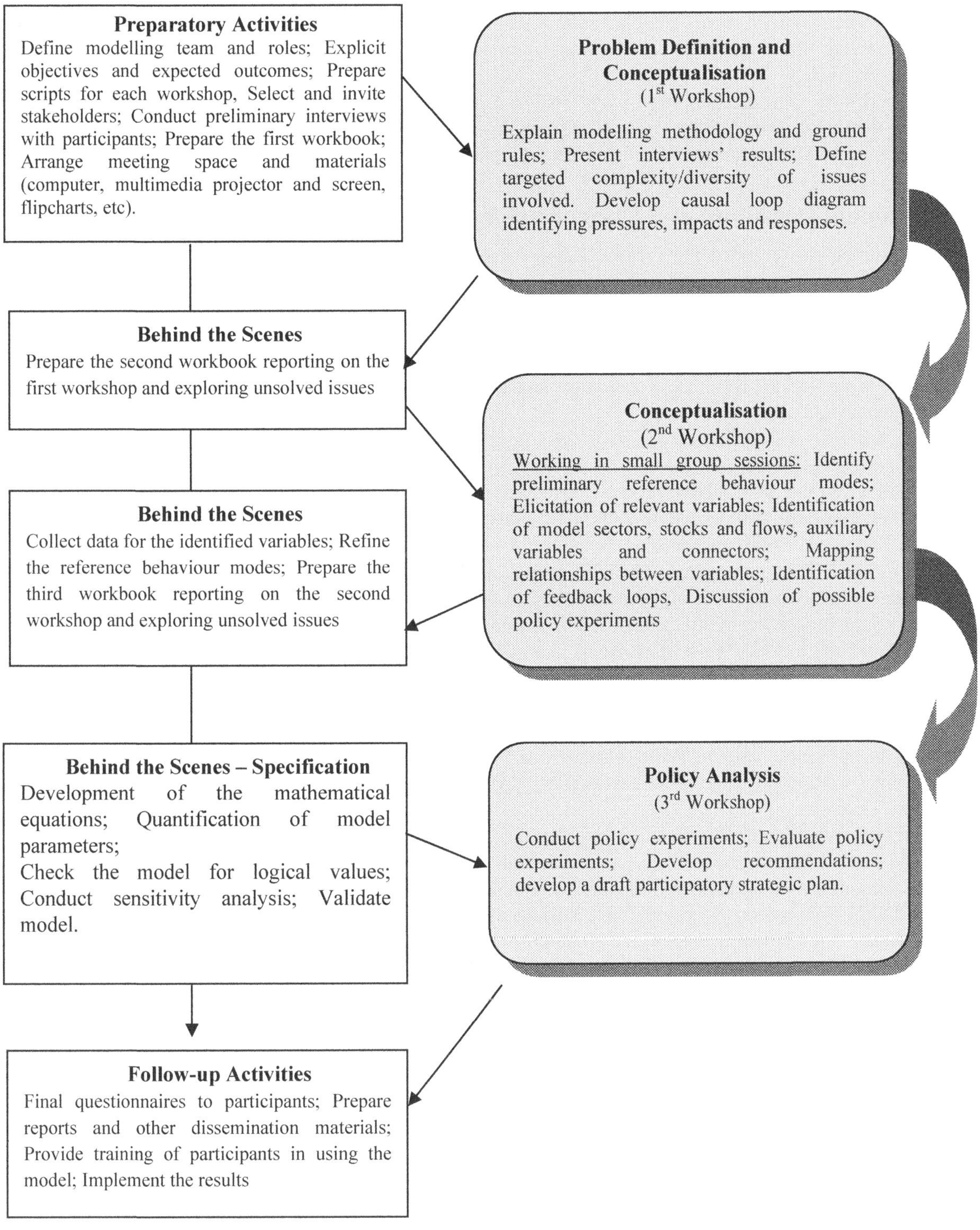

Figure 9.2 Stages of the participatory modelling process in the Baixo Guadiana

During the 3-week period anticipating the workshop, the modelling team has managed to conduct 32 semi-structured interviews with confirmed participants (a request for setting up these interviews was included in the invitation). A reassuring result for the stakeholder analysis was the fact that only two of these interviewees did not consider the invited list as being representative of the stakeholder groups in the Baixo Guadiana.

The interviews were backed up by a questionnaire. Open questions addressed the issues of participation, perceived problems and desired opportunities, which the interviewees wished to explore for the future of the Baixo Guadiana. Other sections presented verbal statements according to which respondents revealed their level of agreement or disagreement. Preliminary interviews were a useful tool for building rapport with the participants and explaining the participatory process in more detail. These interviews also allowed for collecting relevant background information from the involved actors. Results from the interviews represented a point of departure for discussing the issues to be addressed by the models.

The main results from the interviews were analyzed and compiled in a workbook that was distributed at the beginning of the first workshop. This workbook also included information on the participatory methodology, specifying the procedure for developing qualitative models (causal loop diagrams).

9.4.2. Problem Definition and Conceptualisation

Fifty-seven participants attended the first workshop, which was scheduled for half a day. The group included 43 representatives from 29 organizations and 14 property-owners. The represented organizations included the national water authority, two protected areas, regional directorates for the environment, agriculture and economics, local municipalities, regional development associations, environmental associations, producers' associations, research institutes and tourism developers. This provided a mix of decision-makers, technicians, scientists and local citizens. Nevertheless, some relevant organizations did not send any representative to the meeting (e.g. domestic water supply firms and regional Spanish administration).

After the briefing on the overall objectives and implementation process of the WFD, the group discussed the strategic importance of water management in the Baixo Guadiana and the role of participation within this context. Participants referred that an effective level of participatory impact should be able to promote sharing of power and knowledge between those with vested interests in the Baixo Guadiana. There was a strong discontent for "traditional" information and consultation mechanisms, as these do not provide real opportunities for sustaining better-informed decisions. Participants also recognized that the absence of a broad consensus on current views and future visions for the Baixo Guadiana was a consequence of the fact that quality information was not available to all the interested parties.

The workshop proceeded to the core task of sharing and possibly aligning participants' mental models. The distributed workbook included a list of problems identified during the preliminary interviews, which were grouped in six sectors: water quality and quantity, agricultural development, nature conservation, tourism, institutions and social issues. The facilitator planned a piecemeal procedure to conceptualize the causal loop diagrams describing problems, causes, effects and solutions in each sector. Self-stick flipcharts were used to draw the causal relationships between the variables in an interactive fashion. When a sector was complete two members of the team assembled the individual parts on a larger wall, identifying interactions between different sectors to form a holistic view of the current problems of the Baixo Guadiana.

The facilitator asked for the selection of a "problem-variable" when working in each of the sectors. This variable was placed at the centre of the diagram. In the second step, the possible causes of the problem were identified as well as the polarity of the causal relationship between them. For instance, a negative relationship was built between river flow regimes and water salinization. A negative link indicates that the variables change in the opposite direction. All things remaining equal, "an increase in intensive agriculture practices upstream the Baixo Guadiana leads to a decrease in rover flows, which may in turn lead to an increase in the water salinization. The third step consisted of identifying the consequences of problems. Finally, when the participants identified possible responses, feedback loops were drawn linking consequences (impacts) with the causes (pressures).

The qualitative model represents the shared view reached by the group (Figure 9.3). It includes issues associated with water salinization, increase of sediment inputs, risks to nature conservation objectives, and development of sustainable forms of agriculture and tourism. This conceptual model integrated the partial causal loops and formalized the linkages between different problems, which is likely to favor recognition of the underlying feedback processes and interrelationships between sectors (e.g. salinization and intensive forms of tourism both contribute to modification or destruction of habitats). This model also allowed for highlighting possible intervention points addressed at the structural causes of the identified problems.

The information from the qualitative model, complemented with the results from the preliminary interviews, was synthesized in a matrix of Driving forces – Pressures – State – Impact – Responses (EEA, 1999) for the Baixo Guadiana. The encoding of the information in a DPSIR matrix was not explored in detail with the audience to avoid overuse and mix of methods. This was not very relevant for the modelling process. On the other hand it was relevant as a parallel output that could be useful for the river basin authority, in the context of the WFD pilot river basin testing. The DPSIR matrix was made available to the public.

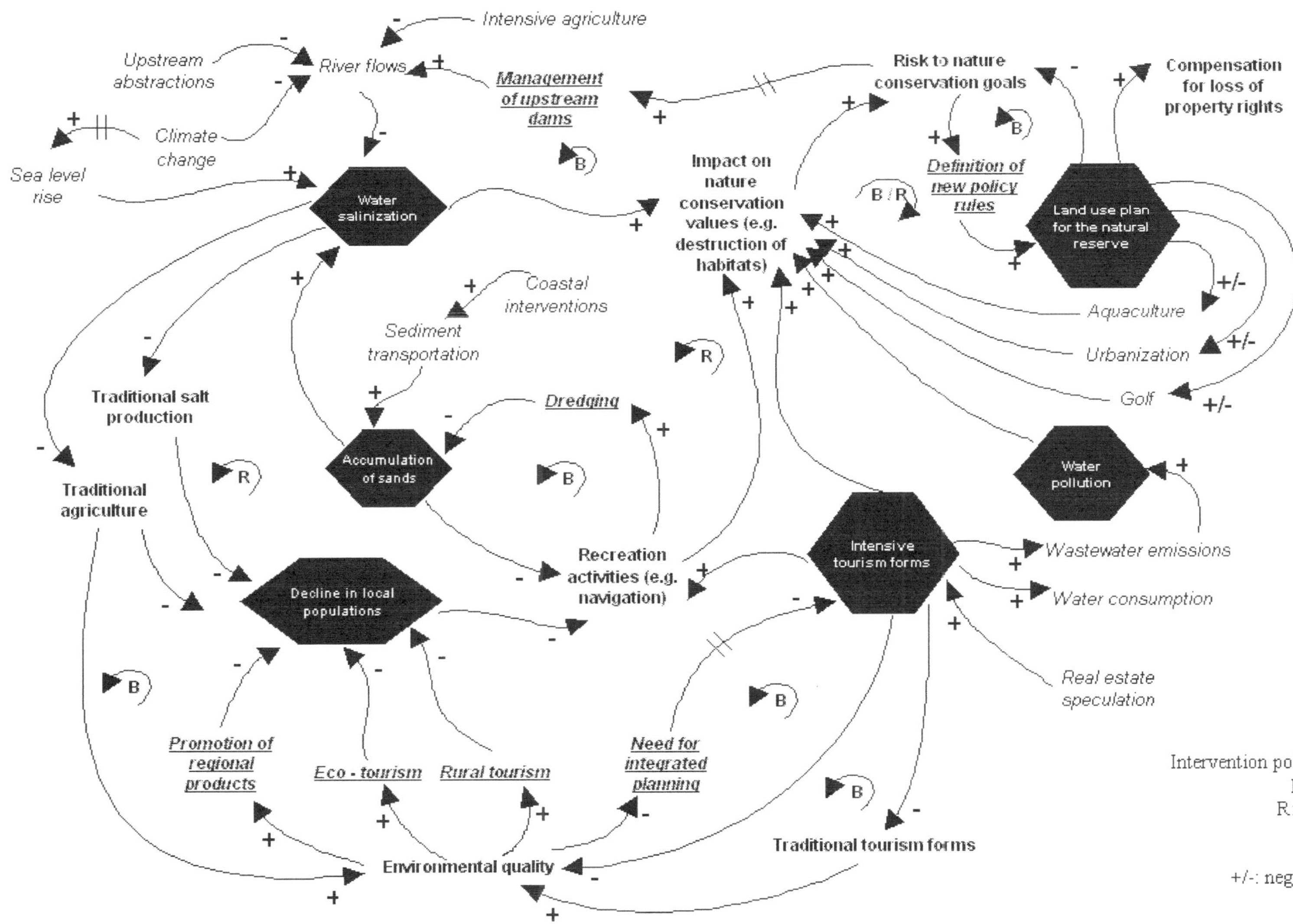

Key to the symbols

Problems: Hexagons
Pressures: *Italic*
Impacts: **Bold**
Intervention points: ***Bold italic underlined***
B: Balancing feedback loop
R: Reinforcing feedback loop
+: positive causal link
-: negative causal link
+/-: negative or positive causal link

Figure 9.3 Qualitative model (causal loop diagram) developed in the first workshop

9.4.3. Conceptualising the quantitative model

The research team met twice again with the gatekeeper and managers of the Natural Reserve for the organization of the second workshop. Invitations were sent to all the 86 stakeholders of the consolidated list. Participants were then invited to work in two smaller groups (working for a half day each), for the conceptualisation of a quantitative model, taking the qualitative model as a point of departure for the analysis.

The morning session was attended by 9 participants, and addressed the issues associated with nature conservation in the Baixo Guadiana. There were representatives from professional associations (salt-making and forests), Natural Reserve managers, a local tourism firm, the Regional Directorate for Economics and INAG. There were last minute cancellations from representatives from other nature conservation organizations. Landowners and local administration did not attend the meeting and provided agenda constraints or no justification at all for this fact.

In the afternoon, 18 participants have attended the second session to conceptualize the model sectors related with water resource issues and sustainable development of economic activities. Adding to the organizations represented in the morning session, participants from the regional development associations, the regional environmental administration, the local administration and two landowners have been present.

The facilitator has repeated the following tasks in each of the sessions:

- Distribution of the second workbook;
- Recalling project objectives, methodology and presenting the objectives for the second workshop;
- Presentation of the summary of results from the first workshop – the list of identified problems, causes, effects, intervention points. Presentation of the integrated causal diagram;
- Conceptualisation: a) Identify variables associated with the problem; b) define time horizon – *How far in the future? How far back in the past lie the roots of the problem? c) Identify reference modes – What is the historical behaviour of key variables and concepts? What might be their behaviour in the future?*;
- Use of flipcharts to list the suggestions from participants for possible information sources for the identified variables. Arrangements were made with participants for the data collection process;
- Asking for potential policies to address the problems (strategic objectives, possible measures and actions/projects). This helped to draft alternative management policies;
- Presentation of the system dynamics quantitative modelling methodology and its building blocks. Examples of stock-and-flow dynamics – "bathtub dynamics". Material and information flows;
- Formulation of the dynamic hypothesis – Develop stock-and-flow maps of the causal structure with the collaboration of the group. The computer screen was projected onto a screen for all participants to follow and directly suggest variables and linkages, with the help of the Powersim software;
- Ending session with a summary of obtained results and prospects for the final workshop.

The conceptualisation exercise was supported by the drawing of the so-called "reference behaviour modes". This is a fundamental step to guide the modelling process and to test the usefulness of the model at the end. For the key variable identified in each sector, the facilitator asked participants about their "observed" behaviour patterns over the reference time horizon (1970- 2004). This allowed for drafting behaviour-over-time plots depicting the perceived behaviour trends, which constitutes a starting point for model construction.

At the end of the workshop, the conceptual stock-and-flow model included a total of 91 variables distributed along three model sectors (Figure 9.4). The dotted arrows in Figure 9.4 represent the interrelationships (defining feedback loops) that were not explicited during the second workshop, but ended up being formalized during the policy analysis stage of the MM process.

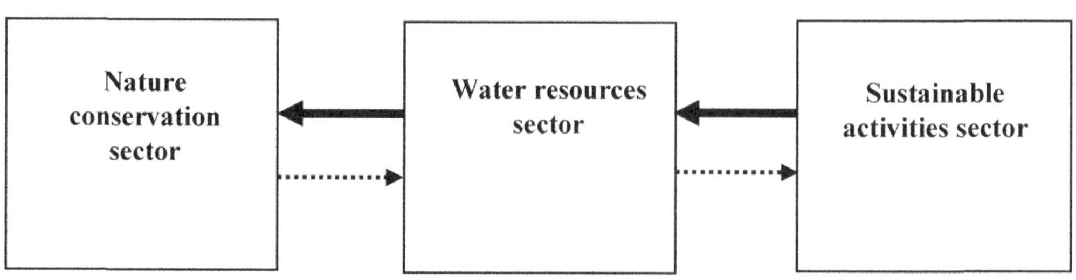

Figure 9.4 Overview of the Baixo Guadiana quantitative model sectors

9.4.4. Specification

In between the second and the third workshops, the research team developed a data collection plan for the specification of the model variables over the reference time horizon. For each sector, several documents (e.g. scientific studies, policy documents) and other information sources suggested by the participants were consulted. Two meetings with nature conservation managers at the Natural Reserve were appointed for specification purposes. In these meetings, the research team organized a round of discussions for quantifying, based on expert knowledge, some variables related with the nature conservation indexes proposed at the second workshop. Several participants were consulted by telephone to provide information they had suggested. Research team members collected data from the Internet, libraries, the National Institute for Statistics, INAG, and regional administration offices. Although some difficulties have been observed (e.g. documents not available for public distribution, delays in receiving the information, suggested sources that did not have the suggested information), it was possible to collect the necessary information to construct a preliminary running version of the model to be used at the final workshop.

A third workbook was produced, explaining the model structure and how it was developed in the second workshop, presenting the elements of the first set of scenarios that could be run (based on suggestions made throughout the whole participatory process – from the preliminary interviews to the second workshop) and indicating the preliminary strategic objectives and measures drafted in the previous workshop.

The modelling team has also checked the model for validation and sensitivity analysis purposes. All variables were documented in the model indicating the sources of information used and the assumptions made. This was demonstrated to participants in the third workshop.

9.4.5. Policy Analysis

A total of 20 participants attended the third and final workshop. There were representatives from the regional development associations, local administration, Natural Reserve managers, regional administration (economics and fisheries), professional associations and tourism firms.

The workshop agenda included the following activities:

- Synthesis of the results from the participatory modelling process, followed by plenary discussion;
- Presentation of the quantitative model, sector by sector, explaining the changes made during specification; showing information sources, and discussing assumptions and ways to improve the model (*e.g.* the sector for the development of sustainable activities was poorly developed, lacking quantitative information for specification of the variables);
- Presentation and simulation of different scenarios;
- Final drafting of the list of strategic objectives and measures foreseen by the group to support integrated and participatory management of the Baixo Guadiana;
- Presentation of follow-up activities and the formal assessment of procedural outcomes (a final questionnaire was delivered to each participant to return expressing the opinion on different topics).

At this policy analysis stage of the participatory modelling process, the structure of the model sectors was presented side-by-side to its behaviour so that participants could observe the fundamental principle of the system dynamics methodology: structure driving the patterns of behaviour of the system. With the use of time-graphs and timetables, it was possible to simulate the model's behaviour from 1970 to 2004 (reference behaviour mode) and then from 2004 to 2015 (scenario analysis).

The scenarios (accepted as different policy experiments) that were analyzed during the workshop were defined considering the suggested projects/measures that have been proposed for the Baixo Guadiana river basin but are still in a planning, evaluation or implementation phase:

- Construction of a wastewater treatment plant serving the municipalities of Vila Real de St.º António and Castro Marim to be located inside the Natural Reserve;
- Licensing of new fish farming industries in the Natural Reserve;
- Licensing of golf courses in the Natural Reserve;
- Restoring the Venta Moínhos saltmarsh in the Natural Reserve;
- Restoring the Taborda saltmarsh in the Natural Reserve;
- Recover traditional salt-making in the Natural Reserve;
- Increasing upstream water abstraction (e.g. for irrigation, urban supply);
- Improving waste water treatment efficiency;
- Regulation of flow regimes according to the Alqueva project.

For each project/measure two alternatives were compared: implementation of the project or measure *versus* no action. The simulation of the alternatives was then observed and the corresponding impacts discussed with the participants.

Some scenarios were particularly interesting since they allowed for creating awareness about the trade-offs of a given intervention. This was the case with the wastewater treatment plant (Figure 9.5).

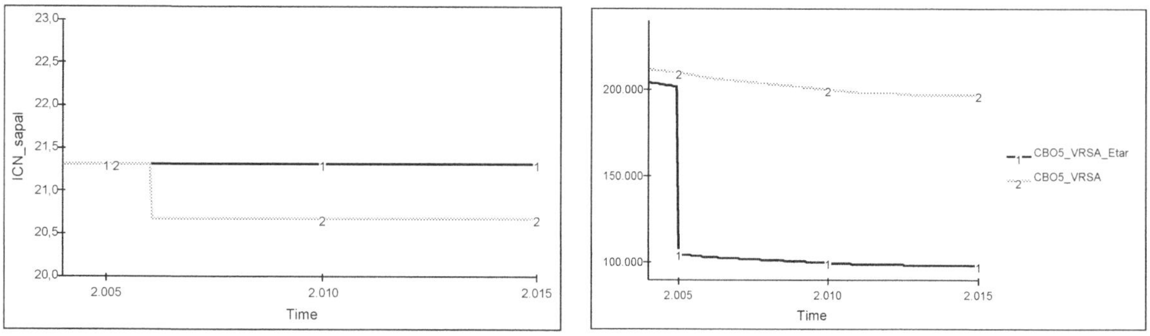

In the nature conservation sector, the construction of the wastewater plant (curve 2) decreases the natural capital index developed for the Natural Reserve

In the water resources sector, the same project will have a positive impact on water quality by decreasing the amount of untreated pollution discharged (curve 1)

Figure 9.5 Example of a policy experiment with the Baixo Guadiana model

If at one hand this project will allow for reducing the annual amount of pollution discharged (measured in the model in terms of Biochemical Oxygen Demand), it will have a negative impact on the natural capital index of the Natural Reserve. This index was expressed as the multiplication of ecosystem quantity by ecosystem quality (both measured as a percentage). Since the construction of the wastewater treatment plant will transform a saltmarsh area in the natural reserve, it will reduce ecosystem quantity, thus decreasing the natural capital index.

9.4.6. Drafting a strategic plan for the Baixo Guadiana

After the policy analysis stage, the participants in the final workshop have collaborated in the synthesis of the strategic objectives and measures for the Baixo Guadiana (Table 9.1). This outcome of the participatory modelling process represents an additional contribution to a sound integrated planning and management of the Baixo Guadiana.

9.4.7. Quality assurance

The Quality Assurance Protocol developed by Corral Quintana and Guimarães Pereira (2004) (see Chapter 5) was used by the research team during the mediated modelling process. Some sections of the protocol were directly incorporated in the design of the workshop sessions. Other items were also incorporated in the final questionnaire, to promote a review by participants of issues related with the quality of information and methods deployed.

The protocol was divided in six parts: framing of the process, social engagement, research group, research procedures undertaken, background information and communication of information. Each of these parts included several topics for analysis, which are described in this section according to the perspective of the MM research team.

a) Framing

Identification of context

The research study was funded under the ADVISOR project by the European Commission, Energy, Environment and Sustainable Development (EESD) Program. The university team carried out the research study, in collaboration (for organization and overall framing purposes) with the Natural Reserve and INAG. The diffusion of results has total access with respect to the main outputs (model, reports, and workbooks available in Portuguese on the Internet).

Framing of issues and identification of alternatives

The definition of issues addressed during the MM process was the responsibility of the participant group. The initial stakeholder analysis has allowed for the identification of a preliminary list of issues according to which participants were invited. These then had the opportunity to suggest others. The invited stakeholders started defining the issues (addressed in the workshops) during the preliminary interviews. These issues were explicitly included in the first workbook and were again discussed and consolidated by the group attending the first workshop. The alternatives identified for addressing the problems discussed emerged from the participatory process.

Table 9.1 Strategic objectives and measures for the Baixo Guadiana proposed by the participants

Strategic objectives	Measures
1.1 Protection of nature conservation habitats and species	-
1.2 Promotion of partnerships and environmental education of local populations	Creation of an association of landowners Continuous information and training of landowners Institutionalization of dialogue
1.3 Settling local populations	Negotiate compatible activities
1.4 Regulation of river flows	-
1.5 Balancing nature conservation with economic development and the welfare of local populations	Integrated Planning Estimating carrying capacities Assess existing capacity for balancing different forms of tourism
2.1 Promote integrated river basin management	Modification of flow regimes Guarantee ecological flows Promote negotiation between water uses Creation of a permanent organization for the integrated management of the Baixo Guadiana
2.2 Maintain good water quality	Provide mechanisms for adequate treatment of wastewater Promote good agriculture practices upstream (especially with respect to intensive irrigation)
2.3 Guarantee navigation in the Guadiana river	Plan for sustainable sand extraction and dredging
3.1 Promotion of sustainable activities	Promote traditional activities (olives, handicraft, cheese, etc) Certification schemes for regional products Biological agriculture Promotion of salt products and traditional salt Promote extensive fish farming
3.2 Promotion of quality tourism	Tourism in the rural environment Link tourism with sports Link tourism with traditional activities Eco-tourism Professional training and certification of traditional fishing methods
3.3 Protect historic and cultural heritage	Dredging for allowing tourist arrivals through the river Create infrastructures Restore traditional buildings with an historic and cultural value

b) Social engagement

Identification of social actors

The process of identification and involvement of stakeholders was described in section 9.4.1.

Effective Participation

The level of participation in the first workshop was of 66% of the invitations sent. In the following workshops this level decreased to around 23%. Although many of the invited participants did not provide a justification for not attending or leaving the process after the 1st workshop (in some cases agenda constraints were referred), the research team makes the hypothesis that the following aspects could have influenced:

- The voluntary effort involved. As explained by van den Belt (2004) in MM research participants are typically not compensated in monetary terms for attending the workshops;
- Given the large extension of the Baixo Guadiana area, social actors were highly dispersed throughout four municipalities. The regional administration is located at more than 60 km distance in Faro. Some landowners in the natural reserve live in Lisbon (300 km away from the natural reserve).
- Given recent experience in these novel MM research processes in Portugal, it was expected that participation would decrease from the first to the following workshops. Usually decision-makers attending the first workshop do not stay for the remaining due to busy agendas and send representatives (at the best). Other participants, although previously informed, attended with the expectation of confronting decision-makers with very specific land use/licensing decisions and although they are given the opportunity to express their opinions,

> they did not have a positive attitude towards collaborating with others, ultimately leaving the process. As stressed by Stave (2002) changes in group composition somewhat limits the potential for stakeholder involvement in the model building process.

To overcome these limitations, the MM process in Baixo Guadiana was planned so that tangible results could be achieved at the end of the first workshop and then at the end of the third event. The larger participant group could first collaborate in the development of a scoping qualitative model (a common view on problems, pressures, impacts and intervention points). The smaller participant group (around 20 participants – the typical number in MM projects) interested in the remaining stages of the process ended up staying for the second and third workshops, collaborating in the development of the quantitative model.

The degree of stakeholder engagement was involvement and collaboration. As described in the previous sections, it is thought that there was a balance of perspectives between stakeholders, public authorities and experts. No clear dominant group was identified.

Identification of perspectives

Plural perspectives were clearly included in the potential list of invited stakeholders and particularly in the first workshop. Despite the decreasing level of participation, it was still possible to have a diversity of perspectives from stakeholders, public authorities, some experts and social actors in all meetings.

c) Research Group

Research Group Experience

Only two MM research projects have been developed in Portugal. The research team in Baixo Guadiana has actively collaborated in both. All members have experience on system dynamics modelling methodology. The perceived experience and flexibility of the research group was reviewed by participants with the application of the final questionnaire. The participants that returned this questionnaire indicated that the credibility of the research team was total (36 %) or high (55 %). They also indicated that researchers conducted the workshops with high (80%) or sufficient flexibility (20 %).

d) Research Procedures undertaken

Moderated Discussion - Modelling

The target of the quantitative modelling process was to achieve a balanced product, made accessible to the general public. All sources of information and methodology were explicitly stated to account for a transparent mediation of information.

Communication of uncertainties was a very important criterion that was explicitly accounted for due to this Protocol. Uncertainties were communicated and written down in the flipcharts. In the second workshop the audience referred uncertainties related with risk. Uncertainties related with the quantification of model variables were accounted for through a rating system developed by the research team where each variable is rated according to the scale 1 (unsubstantiated) to 3 (substantiated).

Citation of other studies was also done, both by the research team and the participants involved. At the end, the final report includes a table indicating the main studies cited and where/who in the MM process has cited them. It should also be stressed that the research team tried to follow all the studies cited by participants to support the specification of the quantitative model.

Encoding qualitative information (DPSIR Approach)

As stated above this output had a very specific purpose of illustrating to the river basin authority how a list of pressures, impacts and responses could be provided in a participatory way. The encoding of the information in a DPSIR matrix was not explored with the audience.

Quantifying information (Participatory modelling process)

Both the qualitative and quantitative models are very relevant for the objectives of the process, as they are the core platform for deliberation. As explained above this method is very flexible in format and content. The model was fully documented as well as the sources of information and the assumptions undertaken. In the workshops there were references to other work apart from the sources stated.

e) Background Information

The process scoped out many problems already studied in the Baixo Guadiana and used information that already existed to characterize these problems. But in some cases the information was generated specifically for this process and in others the information used was previously applied in similar cases (e.g. the nature conservation sector used some of the habitat values used by the Ria Formosa Natural Park managers in the development of the land-use plan of this neighbour protected area).

For the nature conservation and water resources sectors of the quantitative model, the available information was relevant in relation to project objectives. In relation to the sustainable activities sector it was not very relevant. Nevertheless, the uncertainties were recognized. The comprehensiveness of information was average. One of the studies used was confidential, but could be consulted and used for the purposes of this project only. Most of the remaining information was either available without diffusion or in academic sphere.

Because it deals with hydrology processes some of the information used for the water resources sector may only be intelligible to some actors, academics and experts. Most of the remaining information may be easily understood by all the actors involved.

Information was collected using many sources of information:

- Literature, e.g. System Dynamics Review and Ecological Economics;
- Research project reports;
- Internet sites, institutional and non-governmental mainly;
- Experts/scientific consultancy, e.g. managers of the Natural Reserve provided information on the nature conservation value for the avifauna;
- Social research, e.g. interviews with stakeholders

In summary, sources of information included extended facts, historic values, direct calculations provided by the model and estimations. It was concluded that the information available and used partially fulfilled the information needs.

f) Communication of information

The types of information provided to the audience include technical/scientific; social research and policy/regulatory. The information was presented orally and in written documents (e.g. workbooks, reports).

There were clear gaps of information in the sustainable activities sector. The water resources sector had some information gaps; the nature conservation sector was fully accomplished in view of audience requirements. The sources of information and methods were stated. The final deliverables (models and reports) are thought to be balanced products with appropriate illustrations. These products are public – total access, through the Internet.

Major uncertainties were taken into account within the methodology and were communicated to the public (written down and orally).

Pedigree of information was always provided with references, background documents and figures pedigree in the models, presentations and reports.

9.5. CONCLUSIONS

The ADVISOR project was concerned with the development of a framework accompanied by methods and tools for the integrated, participatory evaluation and multi-dimension assessment of projects and plans in the context of the WFD. The design and testing of the implicated set of tools was informed by the lessons drawn from the review of past practice.

The tool presented in this chapter - Mediated Modelling - was used to scope the problems, pressures and impacts perceived by a stakeholder group invited in the Baixo Guadiana river basin in Portugal. The choice of the case study area allowed for raising awareness on the WFD and the need for integrating the objectives defined for protected areas with the environmental objectives established for surface and ground waters. The application of a Quality Assurance Protocol allowed for improving the process of supplying and communicating quantitative information in a participatory modelling process, which is sometimes perceived as a methodological limitation when applying this tool.

The results from the participatory modelling process included a qualitative model (causal diagram), which integrated the perspectives of the 57 participants of the first MM workshop on the main river basin problems (and associated pressures, impacts and responses). Given its simple elements, the qualitative modelling approach seems to be well suited for early problem scoping stages of a planning/evaluation process, especially when there is a large participant group and the presence of decision-makers is desired. Decision-makers usually have busy agendas and will typically not compromise for more time-consuming modelling events, in particular if the exercise is not backed up with strong political and institutional support. Qualitative modelling worked very well in the development of an integrated view and the conceptualisation of the structure of issues. The identification of cause-effect relationships in which system dynamics thinking is grounded also proved to be well suited for the identification of pressures and impacts during the characterization of river basins. This qualitative approach was used in past group model-building interventions but had not been tested before in mediated modelling experiences.

Using the qualitative model as a point of departure, a smaller group of around 20 participants has attended two more workshops to collaborate in the development of a computer simulation model used for experimentation of alternative management scenarios. Quantitative modelling revealed to be well suited for revealing complexity and for making participants more comfortable with it since they contributed in the structuring of the model. As indicated by the final questionnaire responses, the technicalities of the method did not seem to estrange the majority of participants. A simulation model at the scoping level may be further iterated to a research or management model, with increasing detail and applicability in direct decision support in the case study area. The main advantage of quantitative over qualitative models is that only the former may be simulated to compare the behaviour of alternative policy scenarios. Confronting participants with the consequences of different scenarios supports the strategic assessment of alternatives and facilitates the analysis of trade-offs. Although this was not fully achieved in the Baixo Guadiana process, future developments may explore the potential of combining the mediated modelling process with the formal appraisal of the alternatives proposed at the policy analysis stage.

The lessons from the modelling workshops supported the definition of strategic objectives and measures proposed by the group for a sound participatory and integrated management of the river basin. Perhaps even more important than all these outcomes, the participatory modelling approach promoted team learning throughout the structured sequence of deliberative events. This makes participatory modelling a promising tool for application in the development of river basin management plans.

Acknowledgements

The authors would like to acknowledge the support of the Portuguese and the Natural Reserve of the Saltmarsh of Castro Marim and Vila Real de St. ° António. They are particularly appreciative of the support provided by Isabel Pires and Anabela Resende. Furthermore, they would like to thank the collaboration of Serafin Corral Quintana and Ângela Guimarães Pereira in the development and support of application of the specific Quality Assurance Protocol.

REFERENCES

CEC, Council of the European Communities (2000) Council Directive of the European Parliament and of the Council establishing a framework for Community action in the field of water policy. *Official Journal of the European Communities*. L **327**.

ComEC, Commission of the European Communities (2002a) *Guidance on Public Participation in Relation to the Water Framework Directive – Active involvement, consultation and public access to information*, WFD Common Implementation Strategy Working Group 2.9, Brussels.

ComEC, Commission of the European Communities (2002b) Guidance *for the Analysis of Pressures and Impacts in accordance with the Water Framework Directive, Common Implementation Strategy*, IMPRESS Working Group 2.1, Brussels.

Corral Quintana, S. (2000) *Una Metodología Integrada de Exploración de los Procesos de Elaboración de Politicas Públicas*. Universidade de La Laguna, Tenerife. EC - Joint Research Centre: EUR 19724 ES.

Corral Quintana, S., Guimarães Pereira, Â. (2004) Protocol for Quality Assurance of Information and activities undertaken in the Portuguese case study of the ADVISOR Project, WP3 report of the ADVISOR project (EVK1-CT-2000-00074).

EEA, European Environment Agency (1999) *Environmental indicators: Typology and overview*. TNO Centre for Strategy, Technology and Policy: Copenhagen.

Forrester, J.W. (1961) *Industrial Dynamics*. Productivity Press, Portland.

Guimarães Pereira, A. (2003) *Quality of Information in River Basin Governance – Review of case studies: Information Vertex*. Report prepared for the ADVISOR Project, EC -JRC, IPSC Ispra.

Lanz, K., Scheuer, S. (2001) *EEB Handbook on EU Water Policy under the Water Framework Directive*. European Environmental Bureau, Brussels.

Petts, J. (2003) Barriers to deliberative participation in EIA: learning from waste policies, plans and projects. *Journal of Environmental Assessment Policy and Management*, **5**(3), 269-493.

Richardson, G.P., Pugh, A.L. (1981) *Introduction to System Dynamics Modelling with Dynamo*. Productivity Press, Portland.

Stave, K. (2002) Using system dynamics to improve public participation in environmental decisions. *System Dynamics Review*, **18** (2), 139-167.

Sterman, J.D. (2000) *Business Dynamics - Systems Thinking and Modelling for a Complex World*. McGraw-Hill: Boston.

UNCED - United Nations Conference on Environment and Development (1992) *Rio Declaration on Environment and Development, The United Nations Conference on Environment and Development*, Rio de Janeiro 3-14 June 1992.

van den Belt, M., Deutsch, L., Jansson, A. 1996. An Integrated Ecological Economic Assessment of the Coastal Zone of Patagonia: Using Modelling as a Consensus Building Tool. In: *Proceedings of the Conference Ecology, Society, Economy. In Pursuit of Sustainable Development*, Université de Versailles Saint Quentin en Yvelines, Paris.

van den Belt, M. (2004) *Mediated Modelling: A System Dynamics Approach to Environmental Consensus Building* (ed.), Island Press: Washington, D.C.

Vennix, J.(1996) *Group Model-Building: Facilitating Team Learning Using System Dynamics*, John Wiley & Sons: Chichester.

Vennix, J., Andersen, D., Richardson, G., Rohrbaugh, J. (1992) Model-building for Group Decision Support: Issues and Alternatives in Knowledge Elicitation. *European Journal of Operational Research*, **59**, 28-41.

Videira, N., Lobo, G., Antunes, P., Santos, R. (2003a) *Public Participation in Integrated River Basin Governance*. Report prepared for the ADVISOR project (EVK1-CT-2000-00074).

Videira, N., Antunes, P., Santos, R., Gamito, S. (2003b) Participatory modelling in environmental decision-making: the Ria Formosa natural park case study. *Journal of Environmental Assessment Policy and Management* **5** (3), 421-447.

WWAP, World Water Assessment Programme (2003) *Water for People, Water for Life -UN World Water Development Report (WWDR)*, UNESCO Publishing/Berghahn Books, Paris.

WWF, Worldwide Fund for Nature (2001) *Elements of good practice in integrated river basin management: A practical resource for implementing the EU WFD*. WWF/EC Water Seminar Series, Brussels.

Zagonel, A. (2002) Model Conceptualization in Group Model Building: A Review of the Literature Exploring the Tension Between Representing Reality and Negotiating Social Order. In: *Proceedings of the 2002 System Dynamics Society Conference*, Palermo, Italy.

Chapter 10

Social Multi-Criteria Analysis for the Evaluation of Water Management Alternatives. The Case of The Costa Del Sol (Malaga, Spain)

L. Moral Ituarte[1], S. Corral Quintana[2], A. Guimarães Pereira[3], P. Paneque Salgado[4] and B. Pedregal Mateos[1]

[1] Departamento de Geografía Humana, Universidad de Sevilla. C/. María de Padilla, s/n. 41004 Sevilla, Spain
[2] Departamento de Economía de las Instituciones, Estadística Económica y Econometría, Universidad de La Laguna. Campus de Guajara, s/n. 38071 La Laguna, S/C de Tenerife, Spain.
[3] Institute for the Protection and the Security of the Citizen, Joint Research Centre, European Comission. Via E. Fermi, Ispra, Italy.
[4] Departamento de Humanidades, Universidad Pablo de Olavide. Carretera de Utrera Km. 1. 41013 Sevilla, Spain

10.1. INTEGRATED EVALUATION, MULTI-CRITERIA EVALUATION AND THE WATER FRAMEWORK DIRECTIVE

Although the Water Framework Directive (WFD) does not explicitly call for integrated evaluation to implement the several processes that shall govern the water environment of the European Union, it is recognised and also fairly sensible that *integrated* sorts of evaluations have to be performed, to deal with:

(1) the Directive's own wordings of "sustainable activities", "significant effects", "significantly better environmental options", "environmental and socio-economical needs" which call for multi-dimensional analysis;
(2) the complexity and uncertainty associated with available knowledge to govern, plan or foresight the water resources of the EU;
(3) the gaps of scientific and technical knowledge about the water resources and the need to extend the knowledge basis to knowledge other than scientific or technical;
(4) the request of involving relevant social actors in the governance processes (Article 14, etc.), acknowledging a key feature of water governance: plurality of interests, of perspectives, of values and of relationships with water.

The "economics and the environment" guidance document for the implementation of the Directive proposes evaluation tools mostly of economic and cost-effectiveness nature; the document already points out that other expertise and dimensions should be included in assessment processes of water resources. However, it should be understood that the inclusion of other expertise, other knowledge and other dimensions of analysis has to be processed with tools that:

- do not amalgamate those multiple dimensions into a single scale of measurement because they pertain to different actors of decision processes that talk different languages,
- acknowledge the inherent complexity of different types of knowledge,

 Integrated Evaluation for Sustainable River Basin Governance. Edited by Nuno Videira, Giorgos Kallis, Paula Antunes and Rui Santos. ISBN: 1843391481. Published by IWA Publishing, London, UK.

- may account for associated uncertainty and
- since the framework seems to acknowledge extended participation of relevant social actors, evaluation tools should also be able to host shared framing, scoping and assessment of participatory contexts.

Integrated evaluation calls for these types of tools and they are the only way off to responding to the inclusionary spirit of the WFD in some of its parts.

Multi-criteria techniques have been adopted as the means to address environmental management issues. Not only did they potentially allow for conflict analysis and evaluation from multiple perspectives but also embedded an important conceptual stand as far as decision processes are concerned. Roy (1985) pointed out that "the principal aim of multiple-criteria decision aid is not to discover a solution, but to construct or create a set of relations amongst actions that better inform the actors taking part in a decision process".

In the last decade multi-criteria evaluations have evolved, from technocratic to *participatory approaches*, acknowledging the context in which decisions were taking place (Guimarães Pereira & Corral Quintana, 2002). Indeed, not only the algorithmic features have evolved to allow different types of scales to be processed simultaneously, but also the place of multi-criteria evaluation methods has changed. They are used as structuring tools, to support deliberation and decision making processes, assisting on: framing, scoping, generation of alternatives and eventually evaluation and comparison of alternatives. When applying participatory approaches in a cyclical and deliberative way, the decision process becomes a *mutual learning exercise* both for the participating stakeholders and the analysts' team. It upholds the principle that stakeholder participation enriches the evaluation thanks to the multiplicity of perspectives, skills and expertise that it combines.

Nevertheless, participation is a necessary but not a sufficient condition to soundly support and legitimise the decision-making. Taking into account that ethical judgements are intrinsic components of the assessment exercise, conditioning the overall outcome, *transparency concerning assumptions* present throughout the whole process is essential as a guarantee for the study's quality preservation. Participatory techniques are, thus, useful tools for the understanding of the problems but have to be completed by the analyst's responsibility towards society. *Tolerance and plurality* of ethical principles are also essential (Munda, 2002).

10.2. INSTITUTIONAL ANALYSIS, SOCIAL RESEARCH AND THE NAIADE MODEL

10.2.1. Theoretical Background

Understanding the decision making process implies that the social context is properly understood before any decisions are made. The perceptions and positions of the actors who are involved need to be analysed, as well as the relationships that arise among them within the cultural, legal, administrative and political framework of the given case study (Corral Quintana, S. 2000).

The analysis of this complex context, i.e. institutional analysis, is performed in a series of steps. The first one is a thorough study of the proposed problem, which results in its definition and characterisation. The second step includes an analysis of the alternatives and relevant criteria, as well as the relationships among them.

A study of the actors that are involved in the process is a basic issue, closely interlinked with the problem identification. The first step in this regard is to identify who are the main actors, with the aim of making subsequent observations in three aspects: their interests and positions, their capacity to act, and the potential alliances that can take place. The analysis of the social dimensions of decision making processes should allow a better understanding of the effects that can be derived from the adoption of different alternatives, or why one alternative has so far prevailed against all others, perhaps without any apparent reason.

Each actor defines the problem based on his own knowledge, resources and objectives[18]. The fact that actors can have different perceptions allows different problem definitions to occur, sometimes even including opposing views. Stakeholder participation enriches the evaluation thanks to the multiplicity of perspectives, skills and expertise that it combines.

Dente et al. (1998) suggest that the problems inherent in decisions′ contain three fundamental characteristics: its non-objective nature, its instability and its consequences, which will depend on the role of the analyst in the general conduction of the study. On this basis, the analyst team must be the one who "has to define according to his point of view what the central question is in the decision making process" (Dente et al. 1998) and must consider the different

---

[18] A distinction needs to be made between objectives and related concepts such as ideals and wishes. The difference is that an objective goes farther: "the actor has decided to try to make his wish a reality" (Dente et al., 1998).

problem definitions as expressed by the main actors by using his own experience when dealing with these issues (Ingram et al. 1984).

10.2.2. Description of the methodology applied

The first phase of the analysis concerns the *definition of the context and the identification of actors*. Institutional analysis and social research encompassed examination of normative context, multi-scale analysis of media, study of economic political processes and participant observation.

The method applied in the case study bellow presented integrates techniques from multi-criteria analysis, which has its origins in the social sciences and whose use has been recognized as a powerful tool in sociological research (Denzin, 1978). The opportunity of the method is more obvious when it is applied in cases dealing with environmental problems. The combination of social techniques with multicriteria ones improves the understanding of the results of the analysis and clarifies the positions of the actors involved in the process (Corral and Funtowicz, 1998).[19] It is important that different participatory and interaction tools are used at different stage of the study, allowing for continuous testing of assumptions and biases of the research team (Munda, 2002).

There are three sources that are commonly used when defining the problem and identifying the actors and that have to be adapted to the specific conditions of each concrete case study. The first source comes from the multi-scale media analysis. This source of information is quite relevant as far as it provides a broad vision of the process and the different visions, perceptions and arguments of the actors involved in the process. The comparative analysis will reflect political preferences or bias, social interpretation as well as the relative importance of the event at different spatial scales, such as those represented by the local, regional or national levels.

Formal and informal documents related to the case are the second source of information. Thus, an analysis of the current legislation and its trend for a specific period of time allows the institutional framework, that conditions the issue's scope, to be known. A historical review brings the analyst closer to a chronology of the problem under study. This analysis also provides information regarding the behaviour of the agents in aspects such as alliances, changes in objectives, inclusion or and disappearance of actors or changes in the *geometry of power*. Investigations regarding pamphlets, reports and proceedings of meetings undertaken by the actors also must be added to this documentation.

The third source of information, and arguably the most important one, is in-depth interviews, which purpose is to collect direct information and opinions from the actors involved in the process in an unstructured format. These interviews are open, that is, the analyst does not addresses specific concerns of the problem. Generally, a subsequent round of contact with actors are carried out to specifically address: (i) some aspects of the case; (ii) relationships with the rest of the actors: similarities or disagreements in objectives; and (iii) available resources which influence the process. The objective of these contacts are to collect information about the objectives and strategies of the actors, possibly discovering *hidden agendas* and allowing possible alliances among other actors to be investigated.

The second phase of the analysis includes the *exploration of the alternatives* of the case study and the *definition of a set of criteria* which allows these alternatives to be analysed and compared. The alternatives must be represented in the most general manner possible, including a wide range of actions and solutions present in the case under study. The criteria that are chosen for the analysis of the alternatives must be coherent with the case study, including diverse aspects, such as economic, social, ethical and environmental ones, among others.

At this point the information obtained by means of the different social techniques and criteria applied to the alternatives are incorporated into the multi-criteria model. An interesting result of the combination of social techniques with multi-criteria ones is that the multi-criteria analysis looses some of its technocratic character since the analyst draws upon alternatives and criteria obtained by applying a transparent, cyclic and deliberative social research. The multi-criteria analysis allows different alternatives to be explored as a function of the different chosen criteria. The result of this examination will determine a structure of alternatives to be presented to the actors involved for collective and open discussion. This stage is especially recommended, following the focus group methodology, with a view to obtaining feedback, sharing and clarifying the information collected and discussing the results achieved.

In concrete case study that is presented bellow, the selected multi-criteria analysis model was NAIADE (Novel Approach to Imprecise Assessment and Decision Environments), designed by Munda (1995), its software implementation being done at the JRC, Ispra site (JRC, 1996). The NAIADE model specifically offers a series of features which make it adequate for the study of socio-environmental problems through its combination with socio-institutional analysis, as well as avoid some of the criticisms that multi-criteria methods have received:

[19] The combination of participatory and institutional approaches with multi-criteria evaluation was proposed and tested in the VALSE project, specifically in the case study on the evaluation of water management alternatives in the region of Troina in Sicily, which did not seek to provide solutions to existing conflicts, but to progress towards compromise solutions capable of achieving a high degree of consensus among the different stakeholder groups (De Marchi et al., 2000).

- It allows the use of information affected by different types and degrees of uncertainty, such as qualitative, quantitative, precise and fuzzy information, which is of great importance when processing information that is not wholly accurate, reliable, exhaustive and unequivocal.
- NAIADE differs from other multi-criteria methods in that there is no differential weighting of the different criteria used to evaluate the alternatives. The criteria are therefore not prioritised according to whether they are economic, environmental, social or institutional in nature.
- The purpose of the NAIADE model is not to produce an undisputable or "optimum" ranking of alternatives, but to rationalize the problem and provide a framework for communication among the social actors involved.
- NAIADE model allows for two types of mutually enriching evaluations. The first is *multi-criteria analysis* based on the score values assigned to the criteria of each alternative and performed using a matrix (known as the *impact matrix*). The second is an *equity evaluation*, which analyses the value judgements of the stakeholders involved in the evaluation process for each alternative using another matrix (known as the *equity matrix*) and the possible formation of *coalitions* (by representing a dendrogram with groups of stakeholders who defend one of the proposed alternatives).

The environment included in the social analysis allows an improved information structure and visualization of the problem to be appreciated, as well as in the presentation of the objectives and perceptions of the actors involved. This possibility of structuring information offers two main advantages: (i) the analyst can classify and work with the information in an easier way, taking into account that the definition of the context is not easy and consequently hinders the study of the decision making process and (ii) community members such as political decision makers, involved actors in the problem, or the population at large, i.e. the general public, will be allowed to access and understand more easily the situation under study and the positions of those involved, which is important to assist in the negotiating processes aiming at conflicts' resolution (Guimarães Pereira, Â, *et.al.*, 2004).

10.3. EVALUATION OF ALTERNATIVE OPTIONS FOR WATER SUPPLY OF THE COSTA DEL SOL (SPAIN)

10.3.1. The Costa del Sol: Water Issues

10.3.1.1. A largely tourism-based economy

The Costa del Sol is an area that has experienced rapid growth in tourism in recent decades, which has led to a transformation of its socio-demographic and economic structures, mounting pressure on resources and land use restructuring. The resident population has seen several decades of continuous growth, which has quickened even further in recent years: between 1996 and 2001 the resident population (341,401 inhabitants in 2001) grew by 19%, compared with a 2.8% increase in the region of Andalusia as a whole. This sharply rising population trend, coupled with high tourist traffic, which triples the resident population in the summer months and doubles it during the rest of the year, has resulted in increasing saturation, congestion and land overdevelopment. The population density of the Costa del Sol, including both the eastern and western parts, which account for 76% of the province's population, is almost double the provincial average.

From an administrative point of view, the area under study comprises the eleven municipalities of the joint corporation *Mancomunidad de Municipios de la Costa del Sol Occidental* (see figure 10.1). These municipalities roughly coincide with the hydrographic district known as Subsystem 3 (western Costa del Sol) of Exploitation System I (Serranía de Ronda), defined in the Hydrological Plan of the *Confederación Hidrográfica del Sur de España* water authority.

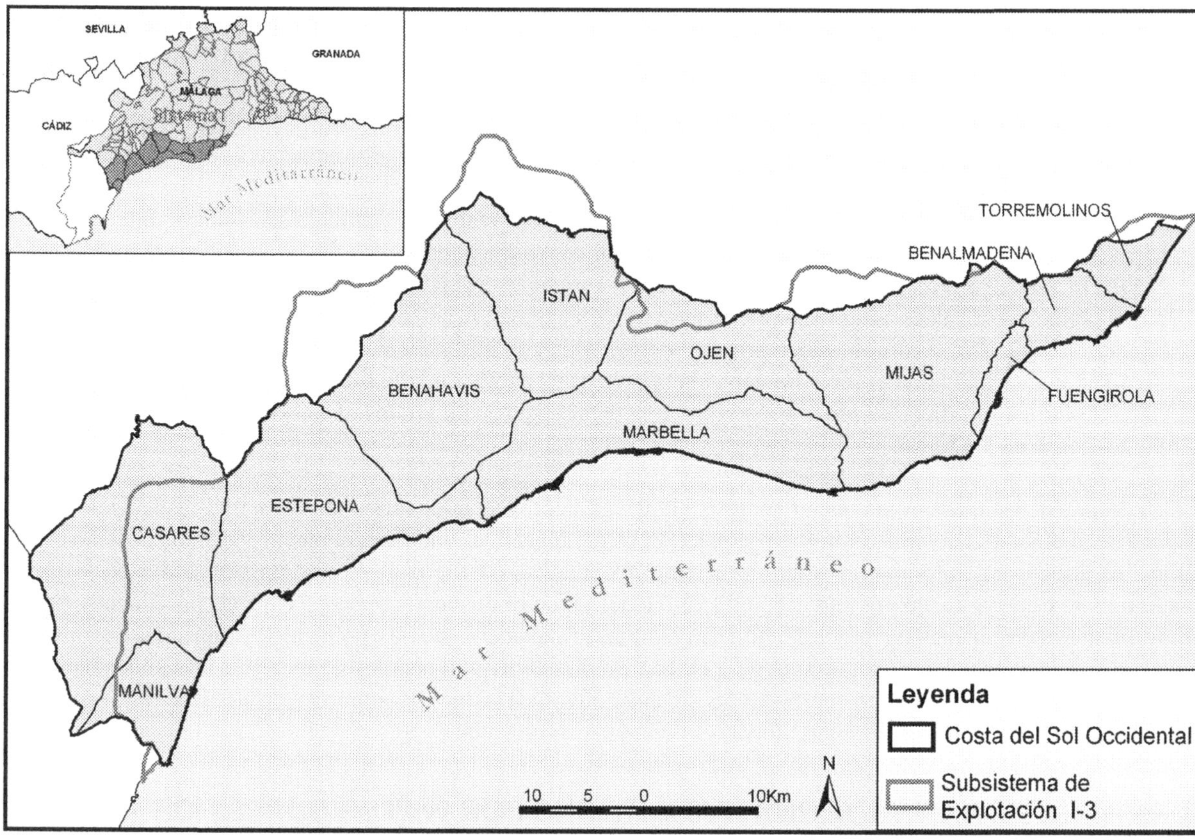

Figure 10.1 Location and boundaries of the area under study
Source: ICA, 1:100,000-scale digital map of Andalusia. Paneque, 2003.

Urban development in the area has been characterised by extraordinarily intense growth in the absence of an adequate regulatory framework. Investment in real estate and the development of the construction sector have been key factors in the case of the Costa del Sol, benefiting from a seemingly bottomless tourism market in Central and Northern European countries. According to data supplied by the Costa del Sol Tourism Observatory (including the western part of the Costa Sol, the area examined in this case study, and the area to the east of the city of Malaga), 8.6 million tourists visited the area in 2003. The average stay was 4.57 days and total spending by tourists amounted to 4,140 million euro.

There exists a development gap between the coastal strip and inland areas which leads to two opposing perspectives on the development trends operating in the area, i.e. one points to the growth dynamics of Malaga's tourism market as a key factor in the overall development of the province and the region, considering territorial polarisation processes as an inevitable consequence. The other point of view stresses the unacceptability of the intense processes of densification and congestion on the coast, accompanied by depopulation, population ageing and the extraction of resources, particularly water, from the inland areas.

10.3.1.2. Hydrological characteristics and water management

Water management strategy in recent decades has been unerringly based on the first of these perspectives. Nevertheless, in recent years, a new logic has emerged in relation to this issue defending the necessity of territorial cohesion.

From a hydrological point of view, the Costa del Sol is an area that is particularly relevant as a case study for the ADVISOR project purposes. On the one hand, it has suffered spells of drought that have highlighted competition and conflict between different water uses, and on the other, the area is characterised by the uncontrolled increase in certain water uses and the lack of anticipatory planning by the competent authorities, whose response has been to instigate reactive emergency measures during drought spells, such as the last one from 1992 to 1995 (Paneque, 2003).

Rainfall in the area under study is relatively high, but varies considerably from one season to another (long and severe summer droughts) and from one year to another (ranging between over 1,200 mm in very wet years and under 300 mm in dry years), owing to the Mediterranean influence (see figure 2). The resulting river regime, also influenced by other physiographic factors (mainly topography, lithology and vegetation), can be described as subtropical Mediterranean, characterised by high water levels in winter and a lengthy period of low water levels. In some cases, the severity of the low flow periods is softened by the abundance of aquifer materials. On the contrary, the fact that the basins of these rivers are relatively small and their courses short and steep, because the mountains where they originate are close to the sea, increases their irregular and torrential regime.

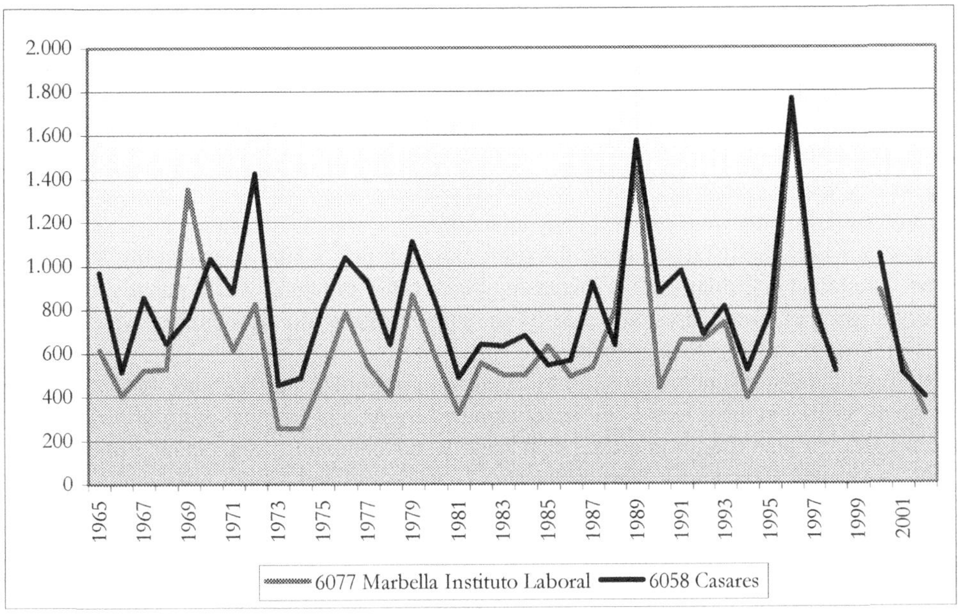

Figure 10.2 Rainfall evolution in two meteorological stations of the Costa del Sol, 1965-2000
Source: Instituto Nacional de Meteorología. Paneque, 2003

These characteristics are a structural conditioning factor in water management in so far as hydrological variability must be adapted to the pattern of social demand. Intervention in the hydrological cycle to control variability and the strong demand for water, inflated by the expansion of highly consumptive, seasonal tourism, are two factors that influence water quantity, quality and environment, defining the backdrop to this issue.

Groundwater plays a significant role. There are numerous hydrogeological units affected by local or seasonal overexploitation, generally coinciding with coastal detritical units, where agricultural and urban demand is concentrated (Fuengirola, Marbella-Estepona). Another serious problem is the deterioration of groundwater quality caused by saltwater intrusion, as a result of excessive pumping of the aquifers. Additionally, there are some hydrogeological units affected by problems of nitrate and pesticide pollution, while half are polluted with heavy metals and three with organic compounds.

The uncontrolled rise in demand and the lack of forward planning have aggravated the impact of drought on the area. The problem is further exacerbated by the conflict between central government authorities and the regional authorities over water management powers and, above all, by the failure of the water authority - *Confederación Hidrográfica del Sur de España*, the River Basin Authority for Southern Spain, under the Ministry of the Environment - to establish adequate channels for public participation[20] (Paneque, 2003).

The organisational structure of water management in Costa del Sol Occidental is particularly complex, further exacerbating existing conflicts. According to Spanish hydrological boundaries, this coastal area belongs to the Southern Spain river basin, which is managed by the above-mentioned *Confederación Hidrográfica del Sur de España*. This authority does not cover a single natural river basin but a number of small basins that empty into the Mediterranean Sea between Gibraltar to the west and the border with the region of Murcia to the east. The *Confederación Hidrográfica del Sur de España* is therefore an administrative hydrographic unit situated on the Mediterranean side of Andalusia region, covering the provinces of Cadiz, Malaga, Granada and Almeria and grouping together a series of sub-basins that are independent of each other from a hydrological point of view.

Closely linked with lack of public participation, another aspect of the debate on the problem of the Costa del Sol water supply and the alternative solutions is the growing complexity and opacity of the management structure. Direct responsibility for the management of urban water supply in the area is shared by the company *Acosol*, which belongs to the Costa del Sol Occidental Joint Municipal Corporation, and by the Town Councils. The first is responsible for

[20] In the final stage of drafting this report, the outcome of the elections held on 14 March 2004 changed this situation. The fact that the Socialist Party is now the governing party both in the central and regional government opens up new prospects for a solution to the conflicts that have dragged on for 20 years. In fact, on 1st January 2005 the competences concerning the Southern Spain river basin were transferred to the Andalusian regional government.

managing the raw or bulk water supply and sewerage services in the eleven municipalities that it covers, while the Town Councils are responsible for local distribution and direct billing of end users. The Town Councils or local authorities can either manage these services directly through a municipally-owned company or charter commercial companies to do so, which makes increasingly difficult to obtain information on basic aspects of water management.

These local enterprises - in addition to distributing water supplied by Acosol from the La Concepción reservoir on the River Verde in Marbella, the main surface-water source in the area-, also manage their own groundwater resources which, although affected by a process of progressive deterioration and depletion, can be managed by them more directly and independently and at a lower cost.

Water consumption by users, system efficiency levels and water prices, on which it is also generally difficult to obtain accurate figures, differ greatly from one company to another.

In general, the costs are not wholly passed on to the user, although prices vary considerably from one water supply Company to another. The effect on billings of the increasing block tariff structure, which penalises higher consumption, is in some cases very slight and, in all cases, softened by the impact of the monthly fixed service charge that all users must pay regardless of how much water they consume.

A significant feature of the water supply system in the area, with important implications for water management is, as mentioned above, the difference in the price of raw water supplied by Acosol and that of groundwater. The first is more expensive (around 0.15 €/m^3) than the groundwater that local water companies can obtain from the aquifers in their respective localities (averaging 0.05 €/m^3). This difference in price has led to the overexploitation and sometimes irreversible deterioration of aquifers in the area.

This problem is even more serious when it comes to treated waste water, also managed by Acosol, which is responsible for sewerage services. In this case, the price rises to around 0.25 €/m^3, and there are no regulations in force requiring the treated water to be used for compatible purposes. This explains why, in spite of the situation of alleged water shortages and scarcity, only around 10% of treated waste water is reused, while cheaper, higher-quality water is used to water large areas, such as golf courses and cropland. In the total area covered by the Confederación Hidrográfica del Sur, as in Spain as a whole, urban uses (domestic, municipal, industrial) account for only a small proportion - between 10% and 20% - of total water consumption, with agricultural uses (irrigation) usually accounting for 80% or more of total demand.

In the specific Costa del Sol area, however, agricultural consumption has been progressively overtaken in recent decades by urban and tourism uses to become a very secondary sector of demand. Nevertheless, as observed above, irrigation still accounts for a significant proportion of total consumption in the river basin area, in some cases subject to the problems of subsidised costs and hydraulic and economic inefficiency that often characterise agricultural uses of water. The reduction of water consumption by agricultural activities and reallocation to urban uses therefore continues to be an issue in the debate on alternative solutions to water supply system problems.

Finally, this overview of the characteristics of the water system in the area under study must include a reference to the seawater desalination plant commissioned by the Town Council of the area's main town (Marbella) just as the lengthy period of drought that lasted from 1992 to 1995 came to an end. Since its completion in 1995, it has not been put into operation, because there has been no reoccurrence of such a drought situation. In spite of the fact that the plant is not in operation, amortisation and maintenance costs still have to be met.

10.3.2. Methodology application: actors, problems, alternatives and evaluation criteria

The application of the social multicriteria methodological framework to this case study had the general objective of finding out whether it is useful in implementing the integrated evaluation and participation objectives established in the WFD. The specific goals were to test the practical feasibility of this technique, in addition to its ability to:

- Facilitate understanding and structuring of problems characterised by multidimensional evaluations.
- Contribute to clarifying the nature of conflicts and creating the conditions necessary to find a meeting point between opposing positions.
- Promote active participation in all the phases of the decision-making process and enable actors to acquire knowledge of each other and their respective positions.
- Promote the formulation of innovative alternatives and make it possible to progress beyond approaches blinkered by routine and inertia.

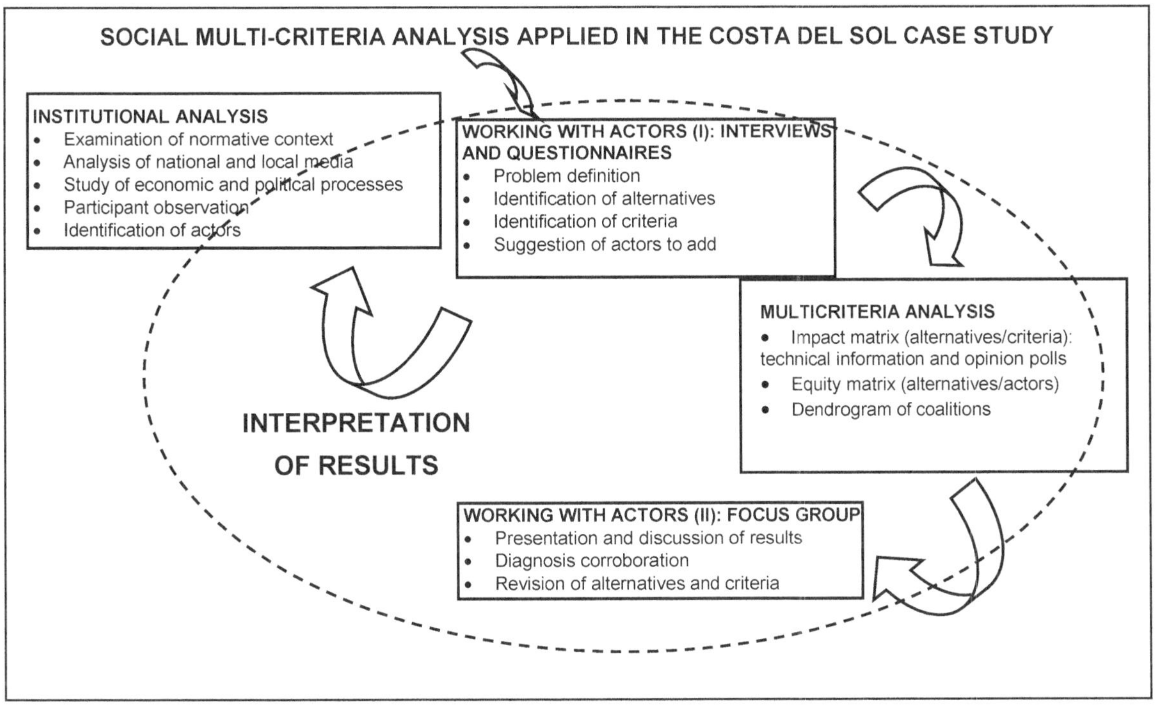

Figure 10.3 Evaluation process' methodology diagram

In order to apply social multi-criteria analysis to the case of the Costa del Sol Occidental water supply, the following stages were implemented (see figure 10.3):

(1) Definition of the problem to be appraised and determination of the scope of study, trough institutional analysis and social research techniques;
(2) Identification of the stakeholders and legitimate interests involved;
(3) Identification of the alternatives and criteria;
(4) Evaluation of alternatives trough multi-criteria analysis;
(5) Presentation of results to the stakeholders involved in the previous research phases, following the *focus group* methodology with a view to obtaining feedback, sharing and clarifying the information collected and discussing the results achieved up to that point.

10.3.2.1. Definition of the problem, determination of the scope of study and identification of the actors and interests involved

Institutional analysis and social research encompassed examination of normative context, analysis of national and local media, study of economic political processes and participant observation – researchers' immersion in the *Foro de Malaga* and the *Agenda 21 of the province of Malaga* – which lead to the definition of the problem and identification of actors.

The time periods selected for the press' analysis were the years 1994-1995 and 2001-2003. The first coincides with a period of drought in the area, when the water debate was particularly heated and there was a barrage of statements made by the actors involved. The second was a more recent period, specifically October 2001 to February 2003, when a further torrent of declarations was prompted by the threat of drought in the autumn of 2002. In addition to the systematic examination of these two periods, certain relevant pieces of information from the period in between were also included. A total of approximately 450 articles from the national daily newspaper *El País* and 800 articles from the local newspaper *El Sur* were studied.

In the stakeholder identification process (see figure 10.4), efforts were made to strike a balance between the number of actors from the public administration (5 participants), business organisations (5), non-governmental organisations (4) and experts (2) to avoid any bias in the weight of opinions and information gathered.

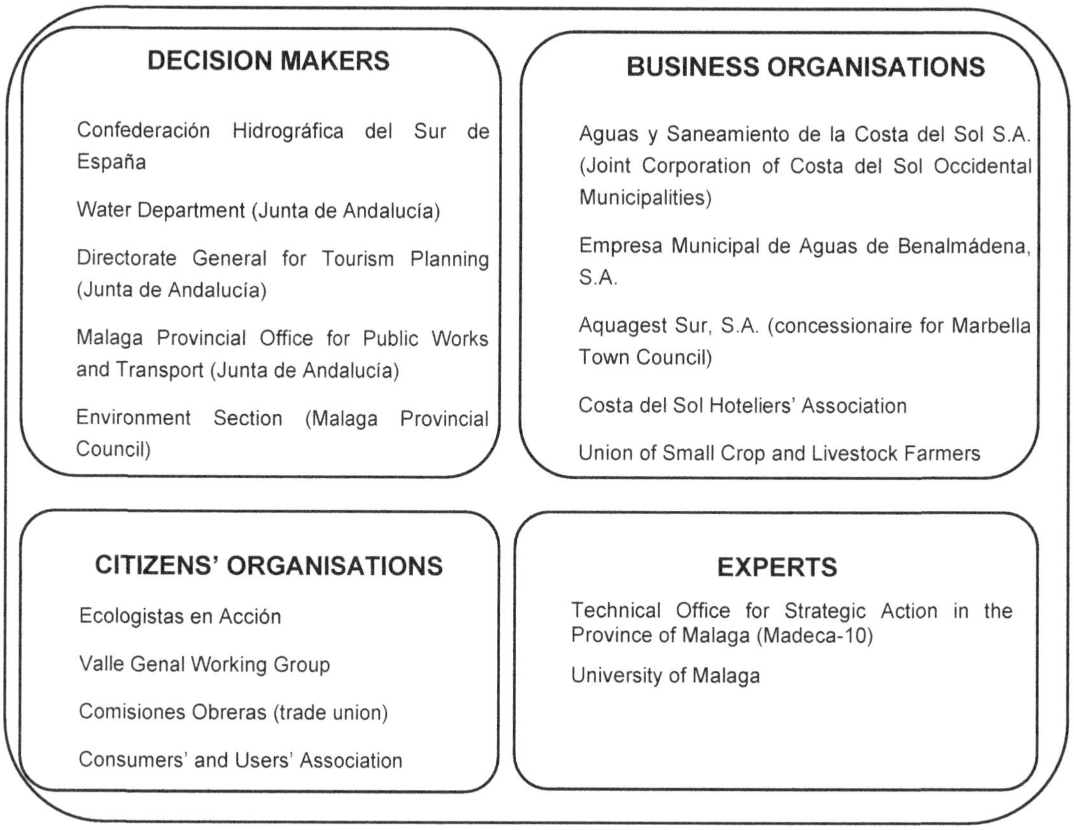

Figure 10.4 Stakeholders identified in the Costa del Sol Occidental area

The objectives of the first phase of the work with the stakeholders were to define or diagnose the problem, identify alternatives and criteria and record suggestions made by the actors. These objectives were fulfilled by conducting in-depth interviews and written questionnaires.

The interviews were conducted according to a set script, which was, however, flexible enough to allow stakeholders to freely express their opinions on any subjects they considered relevant. The interviews were expected to take around an hour, although this was determined by how available the stakeholders were, so that the time ranged between 30 and 100 minutes. All the interviews were recorded with the prior consent of the interviewee, allowing a detailed analysis of their content. The information provided by the sixteen stakeholders in the interviews was clarified and elaborated on in a written questionnaire filled out by them after the interview. Analysis of the interviews revealed that there were gaps in the information required to carry out the multi-criteria evaluation and complete the impact and equity matrixes in the NAIADE model. The written questionnaire was therefore formulated to refine some of the questions already dealt with in the interviews, but which needed to be elaborated on or clarified in some way.

In this way, the stakeholders were involved in the first stages of the evaluation to identify other stakeholders and assist on defining the problem. The problem framing was not intended to be an "expert" description of the current situation of water management in the Costa del Sol Occidental area, but a systematic exposition of the perceptions and knowledge of the stakeholders.

As a result of the interviews and questionnaires, two different perceptions of the water problem in the Costa del Sol Occidental area came to light. One was the definition of the problem from the viewpoint of the authority responsible for water resource management in the area, as reflected in the Cuenca Sur River Basin Hydrological Plan, based on the concept of a 'structural water deficit', affecting the whole of the river basin as well as subbasin units located in the area under study. This diagnosis serves as an argument for investment in new hydraulic infrastructures to improve the water supply in the area, in spite of the fact that at the same time the authority goes to great lengths to downplay the problem in the media, presenting a balanced, problem-free situation. The other stakeholders highlighted the fact that the Costa del Sol Occidental water supply situation is subject to tensions, conflicts and deficiencies, although the main reason for the problem is not absolute scarcity but inadequate management. They put forward the view that the underlying problem is not a 'shortage of water' but 'resource mismanagement'. The majority of the stakeholders regarded the Costa del Sol Occidental as an area with sufficient water resources, but beleaguered by incompetent administration, and criticised the lack of forward planning and land-use management in a geographic area suffering the effects of unbounded growth, a lack of coordination among the authorities responsible for water management and those responsible for territorial and economic planning and management as well as weak, ineffective water management information and participation mechanisms. This does not mean, however, that there are no differences of opinion among them with regard to the need to continue to implement measures to increase water resources by constructing hydraulic infrastructures. Some of them (water supply companies) agreed with the Confederación Hidrográfica that new hydraulic supply-oriented infrastructures are necessary, while others (regional water authorities, conservation groups) were of the opinion that this was not the best solution. However, they did all agree that measures to improve water management were of utmost importance and should be given top priority.

10.3.2.2. Identification of the alternatives and criteria proposed by the stakeholders

Stakeholder involvement was also ensured in the identification of water management alternatives and evaluation criteria (see table 10.1), as it was in the problem definition stage. The fact that the problem was perceived in different ways and the diversity of judgement values and interests at stake enhanced the process implemented to identify alternatives and evaluation criteria, as the criteria exactly matched the different positions taken by the stakeholders consulted. The proposed alternatives that were accepted by the majority of the actors included both actions aimed at increasing available resources and measures to improve demand management. Most of the stakeholders had more trouble identifying criteria to evaluate these alternatives, beyond the usual categorisation into economic, environmental and social criteria. This made us realise that we needed to provide a more studied criteria proposal as a starting point. A list was therefore drawn up and incorporated in the questionnaire, on the basis of which the stakeholders involved were able to add new criteria and accept or reject those proposed. Finally, the technical formulation, this is to say, the parameters that allow NAIADE to compare the different nature of criteria, i.e. preference relations and crossover points, was made by the research team.

Table 10. 1 Alternatives and criteria identified in the Costa del Sol case study

Alternatives:	Criteria:
1. Heightening of the La Concepción dam	1. Implementation costs
2. Use of desalinated water	2. Operating costs
3. Reuse of waste water	3. Effect on employment
4. Modernisation of irrigation systems	4. Effect on economic activity
5. Rationalised use of groundwater	5. Impact on the ecological status of water systems
6. Improved efficiency and water savings in the urban water supply	6. Impact on other ecosystems
7. Territorial policies to control urban development	7. Visual impact on the landscape
8. Non-intervention: maintenance of status quo	8. Degree of institutional difficulty
	9. Degree of social acceptance
	10. Equitable distribution of costs and benefits
	11. Time required to fulfil the established objective

10.3.2.3. Alternatives evaluation trough multi-criteria analysis

All the information gathered in the previous process were used to implement the following stages: construction of *impact matrix*, *equity matrix* and *dendrogram of stakeholder coalitions* using the NAIADE model.

On the basis of the alternatives and criteria identified, the *impact matrix* was constructed with value scores for each of the proposed water management alternatives, according to the eleven evaluation criteria (see figure 5). This matrix was based on data from specialized literature and technical reports, including quantitative, qualitative, crisp and fuzzy values. In order to evaluate the criteria relating to the public acceptance of each of the alternatives, an opinion poll was conducted among the inhabitants of the area. A total of 425 people were polled: 200 in the city of Malaga, 125 in the Costa del Sol Occidental area and 100 in the rest of the province.

The results allowed a comparison of the alternatives and the generation of a ranking according to the selected criteria. Serious problems were encountered when constructing this matrix owing to a lack of information expressed in quantitative and/or accurate terms. In fact, all the criteria, except the implementation costs for each of the alternatives, finally had to be expressed in qualitative terms.

In order to evaluate the proposed alternatives according to the respective positions of the stakeholders, generating a new ranking, the NAIADE model was again used to perform an analysis based on the *equity matrix*, reflecting the qualitative assessment of the alternatives made by the participating stakeholders in the written questionnaires.

Both rankings coincide significantly, underscoring the robustness of the analysis. In the two cases, the demand-management alternatives (saving water in the urban water supply, reuse of waste water) were the highest scoring. The modernisation of irrigation systems also occupied the same position in both rankings, although in the equity analysis, it was not as far behind the top-scoring alternatives. In both rankings, the bottom two alternatives were the same, non-intervention and the heightening the La Concepción dam, that is, the main alternative for the generation of more water resources.

The *equity* analysis also provided information about the potential formation of coalitions among the actors to defend or veto a given alternative. Based on the positions of the stakeholders, a dendrogram of coalitions was produced, representing the formation of possible 'alliances' among stakeholders in the form of a tree diagram, which indicated the degree of divergence. This provided an insight into which alternatives were more likely to be accepted, although the highest ranking alternatives were not necessarily the most feasible. The strongest possible coalition was observed between *Ecologistas en Acción* and the Genal Valley Working Group, two of the actors representing citizens' organisations. A second coalition, albeit with a somewhat lesser degree of similarity, was formed by the University of Malaga and the Environment Section of the Provincial Council.

10.3.2.4. Final Research Stage: Focus Group

In the final stage of the case study, the research team organised a half-day meeting with the actors involved using the focus group methodology. The meeting was held five months after the interview and questionnaire stage had been completed, in a hotel located in the area under study.

The objectives of the final stage of the work were to:

- Introduce the stakeholders involved to each other and discuss the research process and methodology, informing them of the results recorded to date and sharing the information gathered during the previous stages.
- Corroborate and, if necessary, elaborate on and clarify the information obtained from the stakeholders and the conclusions drawn from their analysis: diagnosis of the problem and definition of alternatives and evaluation criteria.
- Provide the stakeholders with the opportunity to modify the information or the analyses made on the basis of the information supplied previously to include new aspects or to elaborate on those already addressed.
- Discuss the results achieved to date.
- Assess the validity of the methodology as an instrument to facilitate dialogue among the stakeholders involved, testing the feasibility of multi-criteria decision aid tools in participatory environments.

For various reasons, some of the stakeholders involved in the study did not take part in the focus group session. Specifically, nine of the sixteen stakeholders who participated in the previous stages of the work attended the meeting.

This is no mere coincidence, but the reflection of a characteristic of the institutional environment in which the experience was carried out: the difficulty of bringing together a broad group of stakeholders with decision-making powers and responsibilities in water management in the area, owing to political and social conflict surrounding this issue. In the Costa del Sol area, the friction materialised during the period in which the study was in progress (2001-2003) as confrontation and lack of dialogue particularly between decision makers from the River Basin Authority and other important social agents. In addition to this conflictive environment specific to the water management system, a more general institutional feature also influenced the experience: the lack of a well-developed organisational culture in the region that would put such information activities and interaction with other agents on the agenda of decision makers from the institutions and entities involved.

The focus group session began with a presentation of the overall objectives of the ADVISOR project and the results of the previous two project work packages. Afterwards, the methodology used involved a brief presentation of the preliminary results of the methodology carried out, divided into sections, given by one of the members of the research team, followed by an open discussion session (confirmation, discrepancies, observations) moderated by the project coordinator. The other two members of the research team acted as observers and secretaries of the meeting. At the end of each discussion session, the moderator summed up the main contributions and points of agreement. The entire proceedings were recorded on audiotape.

After the preliminary presentation, all the stages of the research process, namely identification of participating stakeholders, definition of the problem, selection of alternatives, identification of criteria, evaluation of alternatives and preliminary results, were presented, and the subject was opened to the floor for debate. A lively debate started up immediately and continued throughout the meeting.

Discussion on identification and selection of stakeholders

In the plan drafted by the research team over the two days of preparatory meetings, the identification and selection of stakeholders emerged as an issue that had to be presented and explained, but which would not be the subject of a specific debate. The reason was that the selection process could not be altered at this final stage of the work, making it inopportune the discussion of the matter if an operational outcome were sought. However, one participant (the representative of the regional water administration) criticised the absence from the meeting of decision makers from business organisations and "real (technical) experts in water management". This observation highlighted two important issues: the unavailability of business stakeholders (representing the tourist sector) to take part in the process and the epistemological approach adopted in the study to go beyond the "technical perspective" and conventional scientific knowledge as the priority mechanisms used to solve environmental problems. Although the observation was important because it referred to a key aspect of the exercise, it was not supported by the other stakeholders. The following clarifications and explanations provided by the research team were accepted: a) limit to the number of stakeholders who could be involved in the exercise; b) availability of those who participated in the previous phase; c) the fact that experts were not excluded, as there were technical specialists among the participating stakeholders; d) the aim was to record opinions and judgements beyond those that could be provided by water management experts.

Discussion on diagnosis

A situation diagnosis of the water supply system in the Costa del Sol area was given, based on the points above discussed, i.e. inadequate *management* of available resources, insufficient *coordination* among administrations, weak *participation* mechanisms and lack of control over urban development and insufficient *land-use.*

The situation diagnosis was discussed extensively, although, in general terms, it was largely accepted and openly defended by some of the participants. It was generally acknowledged that there were *tensions* and shortcomings in total water cycle management. There was some resistance to the fact that the diagnosis explicitly questioned the notion of *scarcity*, on which the strategy for the generation of more water resources is based, although it was accepted that an improvement in water management must be accorded top priority.

In addition to a number of important clarifications (the representatives of Acosol, the water supply and sewerage company, underlined that certain infrastructures were lacking, and the representative of the consumers' Federation, FACUA, highlighted the lack of information and participation), several stakeholders addressed a matter of greater significance, namely the need to fit the situation diagnosis of the Costa del Sol water supply and sewerage system (*hydraulic system*) within the *hydrological system* as a whole (quality of the water environment and associated ecosystems). Different arguments were put forward by the representatives of the Genal Valley Working Group and the Provincial Council to support this plea to contextualise the diagnosis of the hydraulic system within the overall hydrological system framework.

This raises the question, among others, of where to draw the boundaries of the problem at hand. Although environmental aspects were taken into account, an essentially hydraulic approach was adopted in the situation diagnosis and identification of alternatives for the total water management cycle (supply and sewerage) in the Costa del Sol area: mechanisms to generate water resources, distribution, treatment, etc. The proposal made during the focus group meeting required the problem to be rescaled, with significant implications for subsequent phases of the debate. As was seen later, one of the new alternatives put forward (reforestation) was related to this new perspective on defining the diagnosis and also raised the question of specifying the problem to a greater or lesser extent, defining its chronological limits and spatial scale.

Discussion on Alternatives

The alternatives were discussed in detail. Some of the stakeholders remarked that the alternatives identified did not respond closely enough to the situation diagnosis, as had been observed in the arguments put forward by some participants in favour of incorporating the general water environment framework and the hydrological cycle as a whole. There was criticism that none of the alternatives sought to improve water quality.

Various stakeholders advocated the addition of the following new alternatives, which were accepted and included in the final proposal of alternatives to be evaluated:

- *Reforestation*, consistent with the call for the hydrological system as a whole to be included in the debate on the situation of the water supply system.
- *'Single authority' for urban hydrological cycle management.* This alternative raises significant implications in the current confrontation (among administrations, among companies) in the area under study, justifying the lengthy debate on this subject.
- Institutionalisation of user negotiation mechanisms for the reallocation of water resources (*water banks*). This is a less controversial proposal than the above, maybe because such mechanisms are not currently involved in the day-to-day wrangling over water resources. The specific functionality of such mechanisms in the area covered by the case study was not detailed, although reference was made to their role in the reallocation of irrigation water to drinking water supply.

Revaluation of alternatives

Before carrying out a new evaluation including the new alternatives that had been proposed and accepted, the evaluation based on stakeholder positions was repeated, including only the judgements of those present at the meeting. The outcome showed little change, in spite of the fact that seven stakeholders who did take part in the interview and questionnaire exercises were absent from the work session. There was some change in the middle section of the ranking, although the top and bottom (highest and lowest scoring alternatives) remained unchanged. This would seem to indicate that the positions of the absentees did not affect the overall spectrum of the group as a whole.

When the new alternatives (*single authority*, *reforestation* and *water banks*) were introduced in the matrix, the ranking changed significantly. One of the new alternatives (*reforestation*), actively advocated in the debate by only one stakeholder (Genal Valley Working Group), although not rejected by any, turned out to be the highest-scoring alternative. The highly positive appraisal of this new alternative signifies a very broad coincidence of views. Below this alternative, the ranking remained essentially unchanged. Alternatives involving the generation of new resources by creating new infrastructures remained at the bottom of the ranking (*heightening of the La Concepción dam* and *desalination*). The rest of the new alternatives (*coordination of authorities* and *water banks*) also finished at the bottom end of the ranking, above the two lowest-scoring alternatives mentioned above.

In view of the results of the new evaluation, a discussion was opened on spatial and time scales of the problem at hand and its solutions. Is the aim to find a 'specific' solution to the problem or to approach the issue from a 'broader' perspective? There was discussion as to whether 'specific' can be considered a synonym of 'short-term' and, consequently, a term that, while ostensibly implying the idea of something operational and realistic, actually denotes a lack of insight and a reductionist approach.

While the matrix based on the judgements of the stakeholders – both the one produced on the basis of the interviews and questionnaires and the new one including the new alternatives – were of interest to the stakeholders, they expressed their dissatisfaction with the fact that the evaluation of alternatives based on criteria was not presented. One stakeholder expressed the view that these "subjective results", while necessary and useful, needed to be supplemented with "technical results", including economic aspects. The research team attributed the failure to present the matrix based on technical evaluations (impact matrix) to the meeting duration. In order to present and explain the impact matrix, the timescale of the meeting would have had to be extended, making it even more difficult for stakeholders to participate. At any event, as mentioned above, although the impact matrix was calculated as accurately as possible, using all the information that the research team was able to collect, and is useful for the purposes of the study as a whole, the level of verification and technical support was insufficient. The risk of presenting it to actors with specific knowledge of the Costa del Sol water system and the area under study was that it could lead to extremely heated discussion and disputes that would distract from rather than add to the effectiveness of the intended focus group communication exercise.

Stakeholders' assessment on the focus group meeting

Finally, positive comments were made on the methodology, alongside calls for effective participation mechanisms and criticism of the opacity of water management authorities who failed to attend the meeting. At any event, the absence of several important stakeholders, influenced and limited the real success of the experience. The representative from the University of Malaga emphasised that information on the results of the previous stages of the study should have been sent to the stakeholders prior to the meeting. This observation is consistent with the interest shown by the participants in the information provided at the meeting about the positions of the other stakeholders with regard to the proposed alternatives, highlighting the need for communication among stakeholders in the area.

10.4. GENERAL OUTCOME: SOCIAL MULTI-CRITERIA ANALYSIS FOR INTEGRATED and PARTICIPATORY EVALUATION OF RIVER BASIN GOVERNANCE

The evaluation of the Costa del Sol case-study's results is done in the context of its usefulness in implementing the integrated evaluation and participation objectives established in the WFD. Specifically, the validity of the methodology applied will be determined by its ability to (i) facilitate understanding and structuring of problems characterised by conflicting evaluations; (ii) contribute to clarifying the nature of conflicts and creating the conditions necessary to find a meeting point between opposing positions; (iii) promote active participation in all the phases of the decision-making process and enable actors to acquire knowledge of each other and their respective positions; (iv) explore innovative alternatives; (v) deal with different types of information that cannot be amalgamated into a single scale of measurement; and (v) stimulate learning processes involving the stakeholders and the research team.

This methodology implies that the stakeholder identification process is inextricably linked to the background study and runs parallel to the problem definition stage, as it is the stakeholders who frame the diagnosis of the problem at hand, which, in turn, influences the selection of stakeholders, in a dynamic and cyclic process of feedback loops among the various steps of the exercise. Different stakeholders will introduce new perspectives in the process, which means that the identification and selection of the participating stakeholders is a key factor in the process and strongly influences the extent to which the results of the evaluation are relevant to the real-world problem under study and, thus, its operational ability to contribute to successful conflict resolution. It is therefore essential that the social actors be identified and selected on the basis of a careful analysis (documents and media analysis, *participant observation*) of the institutional and power-relations framework in which they operate.

In the Costa del Sol case study, the delimitation of the area under consideration, and therefore the *scale of analysis*, was depended on the identification of stakeholders – based on the prior institutional analysis and confirmed and complemented by contributions from the stakeholders themselves – and the alternatives in play. The selection of these stakeholders was not particularly problematic, as there was broad agreement and acknowledgement as to who the relevant actors with legitimate interests in water management in the area are. It should be taken into account that both multi-criteria evaluation and the focus group operate best with a relatively small number of participants, as this prevents the exercise from becoming too complicated.

The selected actors were spokespeople from various organizations, in most cases, the person in charge. The spectrum of organisations involved was very broad, ranging from the top national and regional authorities in water management to local citizens' groups. The problem of the inevitable subjectivity inherent in such an exercise, stemming from the personality of the actors and their individual perceptions and specific knowledge, was overcome thanks to the cyclic nature of the study and open, collective reassessment of the process. The opinion polls conducted in the area under study were used as a means of introducing the perspectives and judgments of the general public, with a view to complementing the diagnosis made by the stakeholders and measuring the degree of public acceptance of the alternatives evaluated in the process.

Regarding working with the actors, one of the most important factors in the Costa del Sol experience was the lack of availability of some of the selected stakeholders in an environment influenced by a weak participatory culture. On the one hand, it proved impossible to involve key actors, such as the Residential Urban Development and Tourism Federation (*Federación de Urbanizadores y Turismo Residencial*), representing the most powerful and conflictive economic sector in the area. On the other, some of the actors who took part in the exercise had limited availability, owing to (i) lack of time and scheduling clashes between some of the proposed tasks and their work commitments and (ii) friction and conflict among some of the actors. These problems led to interviews being cut short, delays in returning the written questionnaires and, in particular, failure to participate in the focus group session.

Finally, it must be taken into account that in an exercise that takes around a year to complete, it is quite likely for there to be changes in the top authorities of the entities selected to take part while the process is still underway. In the case of the Costa del Sol Occidental case study, there was a change of president in the Union of Small Crop and Livestock Farmers (UPA) and the Costa del Sol Hoteliers' Association (AEHCOS), although the former presidents continued to belong to their respective associations, and the head of the Malaga Provincial Office for Public Works and Transport left to work for Malaga City Council. In any case, the actors that were previously involved were invited to assists the focus group session as they did.

10.4.1. Findings regarding problem diagnosis

In the case of the Costa del Sol area, the evaluation processes carried out prior to the Advisor project can be characterised as follows a) they were based on opinions and value judgements underlying and strongly influencing the problem diagnosis; b) they tended to simplify the complexity of ecological and social processes, sidestepping any analysis of the root causes of the problems and possible consequences for natural and social systems; and c) they did not reflect the uncertainty that defines how these systems work.

Taking into account this background, the methodology framework of this case study has proved effective in overcoming these limitations. The open participatory discussion process made it possible to structure the problem, characterised by highly conflicting approaches, and led to new perspectives on the diagnosis of the problem, i.e. the main issue is no longer exclusively or principally a physical shortage of water, but rather poor water resource management, in a context of growing demand. The water debate led to discussion of broader issues relating to development processes (tourist sector, urban development) that are inextricably linked to water issues. Furthermore, when the reflection process became collective (from individual interviews and questionnaires to the collective dynamics of the focus group), the diagnoses became even deeper and more complex.

10.4.2. Findings related to identification of alternatives and criteria

Regarding the identification of alternatives, the number of alternative actions was not limited or preset and, in fact, changed over the dynamic and learning-oriented evaluation process. This promoted the formulation of innovative alternatives and made it possible to progress beyond approaches blinkered by routine and inertia.

The actors proposed a total of twenty alternatives, of which seven were included in the evaluation exercise by the research team (in addition to the "maintenance of the status quo" alternative). The rest were rejected either because they were similar to other proposals that had already been accepted, mainly partial aspects of broader alternatives, or because they did not receive sufficient support from the other actors. The risk of overlooking a relevant proposal was avoided by giving the actors the opportunity, during the focus group meeting, to revise the alternatives that were finally selected and incorporate any others that they agreed should be included. Alternatives added in this way were reforestation, single management authority and water banks.

The identification of criteria was more problematic, probably owing to the complexity of the concept in comparison with that of the alternatives. The initiative undertaken by the research team to formulate a criteria proposal based on the interviews, on which the actors could continue working, proved key to identifying sufficiently precise evaluation criteria in order to make a comparison of the alternatives. Eventually, there were identified eleven criteria of different kinds (economic, environmental, social and institutional), which it was difficult to find quantitative and/or precise information about. Owing to this *technical uncertainty,* that is, the lack of available information, the NAIADE model was considered to be a very suitable evaluation method, as it is capable of processing information affected by different degrees and types of uncertainty (numerical, linguistic and fuzzy variables), a property of the multi-criteria evaluation tool that was particularly useful in this case study.

A further two criteria, namely *degree of public acceptance* and *degree of institutional difficulty*, were also added. With regard to the first, it should be noted that the opinion poll conducted as part of the study was quite significant, because it revealed that the problem was not viewed in the same way by the *stakeholders* as it was by the *general public* and because it provided an insight into the perceptions of people not only in the area under study, but also in the rest of the province, including rural areas. This introduced an original and novel element of *interscalability* in the exercise. The institutional difficulty criteria – defined as the time required to study, process and execute the proposals, the system of subsidies and cost recovery, the number and heterogeneity of entities involved and environmental impact requirements – made the evaluation exercise closer to real conditions, as it took into account the constraints and limitations that would affect the implementation of the alternatives.

10.4.3. Preliminary evaluation of alternatives

The evaluation of alternatives made using NAIADE highlighted the lack of coherence between the diagnosis made and agreed on by the stakeholders throughout the study and the alternatives that were actually predominant. In other words, the solutions advocated by the public authorities to redress the water supply problem in the Costa del Sol area did not coincide with the solutions that best responded to the selected multiple criteria and did not satisfy the stakeholders with legitimate interests in decision making, a fact that they made clear in the participation process. This led the research team to identify *institutional pressure* as a key factor in the situation, resulting from short-term considerations and the ensuing lack of hydrological planning formulated within a reference territorial planning framework. On the contrary, the reflection and learning dynamics set in motion by the participatory multi-criteria methodology revealed a considerable ability to combat *inertia* and *mainstream trends* in the current decision-making system.

With regard to the analysis of coalitions among stakeholders, it should be noted that an examination of existing conflicts revealed a split between the administrations responsible for water management. On the one hand, there was a possible coalition including the Regional Water Department (regional administration), which opposed the management policy of the *Confederación Hidrográfica del Sur* (central administration), as the position held by the latter showed a leaning towards alternatives involving an increase in the supply of water resources. This split cannot, however, be explained solely by political factors. For example, Acosol (Costa del Sol public water supply and sewerage company) held a prominently individualist position, refraining from forming part of any coalition, in spite of its political affinity with the regional administration.

10.4.4. Findings regarding the focus group session

In the focus group meeting, two considerations key to the analysis of this evaluation exercise were detected. First, when the stakeholders were confronted with the final list of alternatives that had been worked on, they suggested the inclusion of other alternatives initially proposed by a small number of stakeholders, but not included by the research team in the final evaluation exercise precisely because they had not enjoyed extensive support. Following an open debate, one of these alternatives (reforestation of catchment's basins) figured among those receiving most support from the participating stakeholders. This shows that the latter phase, in which all stakeholders meet, can significantly change the final result, as issues overlooked during the individual contacts were raised and debated. Second, in this phase of the participatory process, the absence of certain sectors and, indirectly, the conflicts existing among them, were noted. These stakeholders, who had also been reticent to attend the initial individual meetings (interviews, questionnaires), belong mainly to the tourist sector and, more significantly, to the authority responsible for water management in the area, the *Confederación Hidrográfica del Sur*.

10.4.5. Final comments

The evaluation process carried out in the Costa del Sol case study was cyclic and deliberative, becoming a *mutual learning exercise* both for the participating stakeholders and the research team. It upholds the principle that stakeholder participation enriches the evaluation thanks to the multiplicity of perspectives, skills and expertise that it combines. The participants therefore consider themselves co-producers of the knowledge along with the research team.

The main results of the exercise can be synthesised as follows: (i) it contributed to clarifying the nature of conflicts and creating the conditions necessary to construct a dialogue process among the stakeholders more permanently involved and to find a meeting point between opposing positions and (ii) it promoted active participation in all the evaluation's phases and enabled stakeholders to acquire knowledge of each other and their respective positions.

In the joint session held, the stakeholders showed a flexible attitude towards accepting proposals, opinions and judgements offered by the other participants. One remaining question is whether their attitude would be as flexible if they were taking part in a real-world decision-making process.

The main difficulty encountered during the work was the highly conflictive atmosphere surrounding the water management issue. In the case of the Costa del Sol, this water conflictive arena overrides an extraordinary territorial (land use) conflictivity. The shortage of water resources and the competition for them can be attributed to growing demand resulting from rapid, uncontrolled urban and tourism development, in which strong economic, social and political interests are at stake and which is characterised by a lack of transparency and recurrent incidents of irregular conduct. These characteristics, coupled with the institutional framework of the public water administration, explain why it was particularly difficult to gather relevant information. They also explain the limited availability of certain stakeholders to participate in the study, particularly in the final collective reflection phase. The application of this methodology to the Costa del Sol Occidental case study therefore also highlighted the barriers that prevent the effective implementation of this type of evaluation.

In summary, in the case study analysed, the methodological approach proved to be a useful tool specially in structuring the problem at hand in a collective, flexible and cyclic way and eventually dealing with water management conflicts, as it improved the quality and effectiveness of the information interchange and the reflection process. In this way, it can contribute to determining which policy decisions could be best defended before all the stakeholders and the general public, thus reducing the degree of discrepancy and achieving a certain level of consensus. By opening the social debate on water resource allocation and by mapping all the implications, issues and interests involved, it is possible to find alternatives that overcome inertia and look beyond short-term considerations, while rationalising social conflict and resistance.

REFERENCES

Bana e Costa, C. A. (ed.) (1990) *Readings in Multiple Criteria Decision Aid.* Springer-Verlag, Berlin.

Corral, S. and Funtowicz y, S. (1998) Afrontando problemáticas complejas: La planificación y gestión hídrica, *Ecología Política*, **16**, 111-117.

Corral Quintana, S. (2000) *Una metodología integrada de exploración y comprensión de los procesos de elaboración de políticas públicas,* Universidad de La Laguna, España.

De Marchi et al. (2000) Combining Participative and Institutional Approaches with Multicriteria Evaluation. An Empirical Study for Water Issues in Troina, Sicily. *Ecological Economics*, **34**, 267-282.

Dente et al. (1998) A theoretical framework for case study analysis en Dente, B. y J. Ligteringen, *The waste and the backyard.* Kluwer, Dordrecht.

Denzin, N. J. (1978) *The Research Act: A Theoretical Introduction to Sociological Methods.* McGraw-Hill, New York

Guimarães Pereira, Â., Blasques, J., Corral Quintana, S. and Funtowicz, S. (2004) Tools to Inform Debates, Dialogues and Deliberations – The GOUVERNe Project at the JRC. European Commission: EUR 21189 EN. ISBN 92-894-9350-X

Guimarães Pereira, Â. and Corral Quintana, S. (2002) From Technocratic To Participatory Decision Support: responding to the new governance initiatives. *Int. Journal of Geographic Information and Decision Analysis,* **6** (2), 95-107.

Ingram, et al. (1984) Guideline for improves institutional analysis in water resources planning. *Water Resources Research*, **20** (3), 323-334.

JRC (1996) *NAIADE Manual & Tutorial.* Ispra: Joint Research Centre.

Munda, G. (1995) *Multicriteria evaluation in a fuzzy environment. Theory and applications in ecological economics*, Contributions. to Economics Series,Physica-Verlag, Heidelberg.

Munda, G. (2004) Social multi-criteria evaluation: methodological foundations and operational consequences", *European Journal of Operational Research*, vol.**158**/3, 662-677.

Paneque, P. (2003) *Evaluación integrada de la gestión del agua en territorios vulnerables a riesgos hídricos: aplicación a la Costa del Sol Occidental (Málaga).* Doctoral thesis, Department of Humanities, University of Pablo de Olavide.

Roy, B. (1985): *Méthodologie Multicritère d'Aide à la Décision.* Paris: Economica.

Part C

The Way Ahead

Chapter 11

Future Challenges

The perspective stressed throughout the Advisor project is that the shortcomings of past experience in water resources decision-making should not be seen as insurmountable limitations or as evidence against the usefulness of deliberation. Instead, these difficulties provide challenges for improving both the theory and the practice of deliberation. Some of these challenges have been tackled through the research studies presented in the previous chapters. At the end of the ADVISOR project other challenges arise for the future, paving "the way ahead".

Challenge 1 – Improving / combining deliberative tools

The tools tested in the Advisor project are flexible and amenable to several modifications. For example, the duration of a Visioning Workshop (VW) can be increased, or sessions upon which participants deliberate on evaluation criteria can be added to a scenario workshop or a Mediated Modelling process (MM).

There exist also several other deliberative tools that can be modified and innumerable possibilities of combinations and hybrid tools. Some examples: a VW could make use of MM, or vice versa, a MM start with a VW. Visual models (e.g. Geographic Information Systems, etc) can be utilised in VW or MM. VW and MM could be enriched with the use of explicit conflict resolution or consensus building techniques and sessions. Multi-criteria evaluation could be applied upon deliberatively prepared scenarios. The interfaces between Citizens' Juries and Monetary Valuations could be further explored.

All these possible modifications and combinations suggest for a very promising and challenging **research programme**. This programme is not constrained to water resource planning only, but is generally important for environmental management and community planning.

Challenge 2 - Testing IDDPs in practice

The IDDP presented in Chapter 4 is a conceptual synthesis from the experience of application of separate tools in the Advisor project. An IDDP per se however was not tested in the Advisor project. Experimental applications of real, long-term IDDPs in a specific water resource / river basin plan will allow testing the strengths and weaknesses of the process, introduce improvements, and identify specific needs for specific natural, political, cultural contexts. Particular emphasis should be given in looking for the most effective combinations of different tools and different representation / participation practices.

Challenge 3 - Improving the use of scientific information in deliberative processes

The management of scientific information was recognised as a key issue in all the Advisor deliberative experiments. Whereas deliberation and public participation are necessary in principle for addressing scientific complexity and uncertainty, in practice this ideal meets many problems. Participants in the face of uncertainty tend to defer to perceived scientific expertise or opt for simplistic or emotional solutions. Such problems are exacerbated in contexts (such as those of the southern European cases) where there is a notable lack of even very basic data, not to mention alternative sources of information or expertise.

Post-normal science and the normative guidance of Quality Assurance provide the principles for action. The challenge is to experiment more with these principles in actual decision making processes and to find acceptable and effective ways of combining unconstrained deliberation with scientifically informed debate and decisions.

Challenge 4 – Increasing the decision impact of IDDPs

The theory and "science" of deliberation is one step ahead of its actual practice in real decisions. Deliberative processes have yet to make an impact, other than as experiments, as platforms of civic dialogue, education or at best, social learning. In a few instances, they have been used for consultation processes (e.g. in controversial scientific issues). The challenge is how to make deliberative processes an essential component of democratic decision-making. This challenge is even more pronounced for water resource planning and management, which in many cases is still considered as a highly technical issue, constrained to debates between "experts".

Challenge 5 - Changing governance structures

Deliberative processes cannot succeed unless they are embedded in a broader socio-political framework that is itself discursive, defined by deliberation principles and supportive (financially, politically) of deliberative processes. Changing the broader political, social and economic framework towards the ideal of a deliberative (or discursive) democracy requires radical changes which go beyond changes *within* existing political structures[21]. More social science research and analysis is needed in identifying the contours of such a deliberative democracy and its requirements, success and barrier factors and applicability in different socio-cultural or political-historical contexts.

Beyond research however, actual steps and transformations in the political domain are needed. Recent developments in EU policy, such as the White Paper on Governance or in the environmental sphere, the Aarhus Convention and its transposition in Community law, are some first, small steps. More daring steps will be needed in constitutionalising and regulating/institutionalising deliberative democratic processes. In particular clear rules ("entitlements") to deliberate or "represent" should be established and substantial public human and financial resources devoted for deliberation purposes.

The WFD leaves much to be desired in terms of such a shift towards deliberative water governance. There is no clear definition of the level and type of participation expected nor clear entitlements on who has the right to "represent" or "participate". Furthermore, there are questions concerning the real leverage of the Directive in comparison to other policy and power structures (e.g. economic, agricultural policies). Still, it is a first step upon which deliberative applications can build and if successful, gradually contribute to the evolution of broader deliberative water governance.

The Research Challenge

EU Research has an important role to play in supporting the development of IDDPs in water, and more generally environmental, policy. Key areas of research include.

(1) Research and development of modified, new, or hybrid tools.
(2) Demonstration of (and learning from) application in actual decision-making cases, especially in the context of the WFD and other EU environmental policies.
(3) Inter-comparative research on applicability of tools in different regional/local contexts.
(4) Theory of deliberative democracy and discursive governance.

Preferably, research on deliberative processes should be recognised as a separate **key theme** (for a Network of Excellence or an Integrated project), either in the environmental and/or governance related work-programmes, rather than included in other themes. This will help in creating a "**critical mass**" of European research in deliberative processes.

Exchange of information between practitioners of deliberative processes around Europe is urgently needed. There is a communication "void" between practitioners working in similar deliberative tools but in different scientific disciplines, or in different countries. The long and considerable experience with local deliberation practices in development studies is not sufficiently taken up by new researchers in environmental policy studies. There is also a relative lack of exchange of expertise between the significant number of deliberation practitioners in the U.K. and the rest of Europe. EU research **support actions** (including conferences, thematic networks, etc) can provide a useful tool for strengthening such expertise.

Capacity development of future researchers and practitioners is also important. Opportunities for Advanced Study Courses, Marie Curie research grants, research exchanges, etc. should be fully explored.

There are also several opportunities for **International Co-operation** in the field. Experience in deliberative processes, especially from North America and Australia where similar research programmes have been developed, could enrich European efforts. Vice versa, European experience could be tested, adapted and learn from different geographical or cultural contexts.

[21] Dryzek, J.S. (2000), Deliberative Democracy and Beyond. Liberals, Critics, Contestations. New York: Oxford University Press.

Annexes

ANNEX I

Protocol for Information used in ADVISOR (Mediated Modelling case study)

FRAMING

The way in which the *problematique* is framed determines how the participatory activity, the contract with participants, and the outcomes are planned.

Characterisation

Identification of context

Purpose of the activity

Research on the Issue	Testing tools or methodologies	Planning/Decision Support	Public Awareness	Other

Study funded by ...

Supranational/National Funds	Regional/Local Authorities	Private Investors	NGOs	Other

Study carried out by ...

University/Research Centre	Consultancy	Public Officer	Other

Diffusion of results

Total Access	Public Authorities	Academic Sphere	NGOs/Private Agent	Other

Quality Review &/or Extended Quality Review

Framing issues (problematique, governance issues, etc.)

Definition of issues - How

Institutional Analysis	Social research	Expert consultation	Political statements	Other

Definition of issues - Who

Actors	Experts	Public Authorities	Research team	Other

Encoding (for definition purposes)

Addressed as defined	Similar cases	Quantification of objectives	Other	Not known

Identification of alternatives

Definition of alternatives - How

Participatory processes	Experts consultation	Public Authorities	Research team	Other

Encoding (for evaluation purposes)

Addressed as defined	Similar cases	Proxy through quantification	Other	Not known

SOCIAL ENGAGEMENT

Characterisation

Identification of social actors

Identification - how

Institutional Analysis	Experts' opinion	Actors' suggestion	Research team's decision	Public Authorities' decision

INVOLVEMENT - WHO

☐ Experts
☐ Stakeholders
☐ Public Authorities
☐ NGOs/ Civil Organisations
☐ Citizenship

Effective participation

Level of participation (% of invitations sent)

More than 80%	60% - 80%	40% - 60%	Less than 40%

Balance of perspectives (the dominant position)

☐ Stakeholders
☐ Public Authorities
☐ Experts
☐ Research group
☐ Other, please specify ______________________________

Degree of engagement

Empowerment	Collaboration	Involvement	Consultation	Information

Quality Review &/or Extended Quality Review

Identification of perspectives

Acknowledgement of plurality of perspectives

Aims at knowing who determines the characteristics of the planning issue. Involving people is not enough but is a necessary condition to assure the inclusion of different perspectives

Majority of social actors	Main Stakeholders	Public authorities	Experts	Research group

RESEARCH TEAM'S ROLE

Quality Review &/or Extended Quality Review

Research group Experience

(to review framing & structuring)

Competence

Experience[i]:.

High	Medium	Low	None	Unknown

Flexibility[ii]:

Fast Adaptability	Enough	Low	Unknown

Method[iii]:

Institutional Analysis	Stakeholders' Opinion	Independently	Similar Studies	Political Sphere

Legitimacy

Verification.[iv]

External	Independent	Regular	Research team	None / Unknown

Colleague Consensus.[v]

Complete	Some Discrepancies	Competing Schools	Not Enough Knowledge	Unknown

Extended Peer Acceptance.[vi]

Total	High	Medium	Low	None

RESEARCH PROCEDURES UNDERTAKEN

Quality Review &/or Extended Quality Review

Moderated Discussion - Modelling

MEDIATION OF INFORMATION

Intelligibility

Compliance with target audience adapted to the **timings** of the audience - is there an overview/ summary provided? Adapted to the skills of the audience (**format**)?

Balanced product with illustrations where appropriate and summaries of main points	Academic experts sphere only	Cryptic writing – Only for selected users	Unintelligible

Accessibility of product to the audience

Public - available in the INTERNET or a public body	Public - available upon request	Restricted access	Not accessible

Transparency

Sources of information stated (reliability, etc.)	Methodologies used stated	No information on sources & methodologies is provided

Communication of Uncertainties[1]

Are major uncertainties taken into account?

In the problem formulation	Within the methodologies	Communicated to the audience	No account for uncertainties

How were uncertainties stated?

Have implications for policy and for social context been stated?

YES	NO

Were uncertainty relations with risk (namely consequences for different risk management strategies) been stated?

YES	NO

Is citation of other similar studies done?

YES	NO

Collection and Use of Information

Reliability

Sources of information - WHERE (information providers). What sources have been used? – Source as place where information was available. Categories:

Literature	Peer reviewed journals Newspapers or magazines
Research Project reports	
Publications	Governmental International institutions Non-governmental organisations
INTERNET sites	Institutional Non-governmental Commercial Educational Personal
Expert/scientific consultancy	
Social research	Interviews with stakeholders Group meetings

Sources of information - Method of generation[vii] categories:

Experimentation	Extended facts	Historic Values	Calculations	Estimations

Sources of information - Generators[viii] categories:

Experts	Public Sphere	Consultancy	Stakeholders' Opinions	Population's perception

Verification[ix] categories:

External	Independent	Regular	Occasional	None / Unknown

Colleague consensus[x] categories:

Complete	Some Discrepancies	Competing Schools	Not Enough Knowledge	Unknown

Extended peer acceptance[xi] categories:

Total	High	Medium	Low	None

Sources of **Legitimacy**[xii] categories:

Academic Sphere	Public Sphere	Religion	Mass-media	Perception (Traditional Heritage)

[1] This section is based on Guimarães Pereira & Corral Quintana (2001).

Has the information available fulfilled the information needs?

Very well	Partially	Not at all	Unknown

Has the information used (from the available) fulfilled the information needs?

Very well	Partially	Not at all	Unknown

Encoding Qualitative Information – Actors' information (DPSIR approach)

Fitness for purpose

Adequacy/ Applicability - Are the models deployed adequate to the *problematique*? Were the models used fully applicable to the work done?

Specific	Applied in Similar Cases	Generic to the issue	Other (state)	Generic to all sorts of problems

Relevance of models with regards to the objectives of the project - categories:

Very Relevant	Relevant	Sufficient	Not very Relevant	Irrelevant

Adaptability/Flexibility – Could the models be adapted/ tuned to the questions to be addressed?

Very flexible	Flexible	Sufficient	Inflexible

Transformation/ encoding – What kinds of information 'transformations' are performed through the models?

None	Simplifications by highlighting 'key' aspects[2]	Aggregations	Extrapolations	Total

Transparency

Models' documentation - Are the models used documented?

Fully	Partially	Not at all	Unknown

Sources of information - Are the sources of information used clearly stated?

Fully	Partially	Not at all	Unknown

Apart from the sources stated are there **references to other work** that could be relevant to the overall application?

YES	NO

Arbitrariness - scientific set-up: Are assumptions clearly stated while applying the models?

Assumptions fully documented	Assumptions partially documented	No	Unknown

Legitimacy

Colleague consensus categories for methodologies used:

Complete	Some Discrepancies	Competing Schools	Not Enough Knowledge	No Opinion

[2] In this case the reasoning for KEY aspects should be provided.

Quantifying Information (STELLA process)

Fitness for purpose

Adequacy/ Applicability - Are the models deployed adequate to the *problematique*? Were the models used fully applicable to the work done?

Specific	Applied in Similar Cases	Generic to the issue	Other (state)	Generic to all sorts of problems

Relevance of models with regards to the objectives of the project - categories:

Very Relevant	Relevant	Sufficient	Not very Relevant	Irrelevant

Adaptability/Flexibility – Could the models be adapted/ tuned to the questions to be addressed?

Very flexible	Flexible	Sufficient	Inflexible

Transformation/ encoding – What kinds of information 'transformations' are performed through the models?

None	Simplifications by highlighting 'key' aspects[3]	Aggregations	Extrapolations	Total

Transparency

Models' documentation - Are the models used documented?

Fully	Partially	Not at all	Unknown

Sources of information - Are the sources of information used clearly stated?

Fully	Partially	Not at all	Unknown

Apart from the sources stated are there **references to other work** that could be relevant to the overall application?

YES	NO

Arbitrariness - scientific set-up: Are assumptions clearly stated while applying the models?

Assumptions fully documented	Assumptions partially documented	No	Unknown

Legitimacy

Colleague consensus categories for methodologies used:

Complete	Some Discrepancies	Competing Schools	Not Enough Knowledge	No Opinion

1 BACKGROUND INFORMATION (Scientific & Technical Information)

Quality Review &/or Extended Quality Review

Fitness for purpose[xiii]

(a) Correspondence between information used and problem – **Adequacy, Relevance**[xiv]

Adequacy of information categories:

Specific	Applied in Similar Cases	Generic to the issue	Other (state)	Generic to all sorts of problems

Relevance of information in relation to the objectives of the project - categories:

Very	Relevant	Sufficient	Not very Relevant	Irrelevant

[3] In this case the reasoning for KEY aspects should be provided.

Relevant				

(b) **Accuracy** of information - are uncertainties clearly stated? - Categories:

Highly accurate	Accurate Uncertainties recognised	Accuracy not known: Uncertainties not stated	Not accurate	Unknown

(c) **Comprehensiveness/completeness**[xv] of information categories:

Total	High	Average	Low	Unknown

Applicability

(a) **Access & Availability**[xvi]: Was the information available? Was the information accessible (even if available)?

Access of information categories:

Total Access	Without diffusion	Academic Sphere	Not public: need for permits & credentials	Unknown

Availability of information categories:

Available and publicised	Available: without diffusion	Available only at the academic Sphere	Not available	Unknown

(b) **Intelligibility**[xvii] Is information legible for those carrying out the work of the project (or *who can understand the information collected or yet which audiences was the information targeted to*?)

Intelligibility of the information categories:

Actors Involved	Public Sphere	Expert Consultancy	Academic Sphere	Experts

Reliability

The evaluation of the Reliability of the information ranges from its internal aspects (such as its generation and verification) to those more external, such as those related to the acceptance of the specific information by the relevant community.

(a) **Control** (Sources, verification, colleague consensus)

Sources of information - WHERE (information providers). What sources have been used? – Source as place where information was available. Categories:

Literature	Peer reviewed journals Newspapers or magazines
Research Project reports	
Publications	Governmental International institutions Non-governmental organisations
INTERNET sites	Institutional Non-governmental Commercial Educational Personal
Expert/scientific consultancy	
Social research	Interviews with stakeholders Group meetings

Sources of information - Method of generation[xviii] categories:

Experimentation	Extended facts	Historic Values	Calculations	Estimations

Sources of information - Generators[xix] categories:

Experts	Public Sphere	Consultancy	Stakeholder Opinions	Perceptions of population

Verification[xx] categories:

External	Independent	Regular	Occasional	None / Unknown

Colleague consensus[xxi] categories:

Complete	Some Discrepancies	Competing Schools	Not Enough Knowledge	Unknown

(b) **Confidence** (Extended peer acceptance, legitimacy)
Extended peer acceptance[xxii] categories:

Total	High	Medium	Low	None

Sources of **Legitimacy**[xxiii] categories:

Academic Sphere	Public Sphere	Religion	Mass-media	Perception (Traditional Heritage)

Finally:
Has the information available fulfilled the information needs?

Very well	Partially	Not at all	Unknown

Has the information used (from the available) fulfilled the information needs?

Very well	Partially	Not at all	Unknown

ANNEX II

Protocol for Information used in river basin assessment studies

2 Framing

Fitness for purpose[xxiv]

(a) Correspondence between information used and problem – Adequacy, Relevancy[xxv]

Adequacy and Relevance of information categories:

Specific	Applied in Similar Cases	Generic to the issue	Other (state)	Generic to all sorts of problems	Unknown

(b) **Accuracy** of information - are uncertainties clearly stated? - categories:

Highly accurate	Accurate Uncertainties recognised	Accuracy not known: Uncertainties not stated	Not accurate	Unknown

(c) **Comprehensiveness/completeness**[xxvi] of information categories:

Total	High	Average	Low	Unknown

Applicability

(a) Access & Availability[xxvii]: Was the information available? Was the information accessible (even if available)?

Access and **Availability** of information categories:

Total Access	Without diffusion	Academic Sphere	Not public: need for permits & credentials	Unknown

(b) Intelligibility[xxviii] - Is information legible for those carrying out the work of the project (or *who can understand the information collected or yet which audiences was the information targeted to*?)

Intelligibility of the information categories:

Actors Involved	Public Sphere	Expert Consultancy	Academic Sphere	Experts	Unknown

Reliability

The evaluation of the Reliability of the information ranges from its internal aspects (such as its generation and verification) to those more external, such as those related to the acceptance of the specific information by the relevant community.

(a) **Control** (Sources[4], verification, colleague consensus)

Sources of information - WHERE (information providers). What sources have been used? – Source as place where information was available. Categories:

Literature	Peer reviewed journals Newspapers or magazines
Research Project reports	
Publications	Governmental International institutions Non-governmental organisations
INTERNET sites	Institutional Non-governmental Commercial Educational Personal
Expert/scientific consultancy	
Social research	Interviews with stakeholders Group meetings

Sources of information - Method of generation[xxix] categories:

Experimentation	Extended facts	Historic Values	Calculations	Estimations	Unknown

Sources of information - Generators[xxx] categories:

Experts	Public Sphere	Consultancy	Stakeholder Opinions	Perceptions of population	Unknown

Verification[xxxi] categories:

External	Independent	Regular	Occasional	None/Unknown

Colleague consensus[xxxii] categories:

Complete	Some Discrepancies	Competing Schools	Not Enough Knowledge	Unknown

(b) **Confidence** (Extended peer acceptance, legitimacy)

Extended peer acceptance[xxxiii] categories:

Total	High	Medium	Low	None	Unknown

Sources of **Legitimacy**[xxxiv] categories:

Academic Sphere	Public Sphere	Religion	Mass-media	Perception (Traditional Heritage)	Unknown

3 Activities Undertaken

Fitness for purpose

Adequacy/ Applicability - Are the methodologies used appropriate to the work? Were the methodologies fully applicable to the work done?

---

[4] Assumption: presumably the information types used are solely 'qualitative' since either they are textual or derived from oral interviews.

Specific	Applied in Similar Cases	Generic to the issue	Other (state)	Generic to all sorts of problems	Unknown

Relevancy of methodology in regards to the objectives of the project: categories:

Very Relevant	Relevant	Sufficient	Not very Relevant	Irrelevant	Unknown

Adaptability/Flexibility – Could the methodologies be adapted/ tuned to the assessment requirements?

Very flexible	Flexible	Sufficient	Inflexible	Unknown

Transformation/ encoding – What kinds of information 'transformations' are performed through the methodology?

None	Simplifications by highlighting 'key' aspects[5]	Aggregations	Extrapolations	Total	Unknown

Transparency

Methodology documentation - Are the methodologies documented?

Fully	Partially	Not at all	Unknown

Sources of information - Are the sources of information used clearly stated?

Fully	Partially	Not at all	Unknown

Apart from the sources stated are there **references to other work** that could be relevant to the overall assessment?

YES	NO	Unknown

Arbitrariness - scientific set-up: Are assumptions clearly stated while applying the methodologies?

Assumptions fully documented	Assumptions partially documented	No	Unknown

Legitimacy

Colleague consensus categories for methodologies used:

Complete	Some Discrepancies	Competing Schools	Not Enough Knowledge	No Opinion	Unknown

4 Style of Communication

Intelligibility

Compliance with target audience (see also criteria for communication in section 1) Adapted to the **timings** of the audience - is there an overview/ summary provided? Adapted to the skills of the audience (**format**)?

Balanced report with illustrations where appropriate and summaries of main points	Academic experts sphere only	Cryptic writing – Only for selected readers	Impossible to pass from the first paragraph	Unknown

Accessibility of document to the audience

Public document available in the INTERNET or a public body	Public document available upon request	Restricted access	Not accessible	Unknown

[5] In this case the reasoning for KEY aspects should be provided.

Transparency

Sources of the information stated (reliability, etc.)	Methodologies used stated	No information on sources & methodologies is provided	Unknown

Uncertainties

Are major uncertainties taken into account?

In the problem formulation	Within the methodologies	Communicated to the audience	No account for uncertainties	Unknown

Are implications for the decisions stated?

Precautionary measures	Go ahead despite uncertainties	Suggestion for further studies	Not an issue	Unknown

Is pedigree of information provided?

YES	NO	Unknown

ANNEX III

Definitions of Criteria used in Quality Assurance Protocols

---

[i] Experience. skill of the analyst on dealing with similar issues to the one at hand. It is important factor since greater practice in the treatment of similar issuers should imply a greater degree of precision and reliability

[ii] Flexibility: analyst's adaptability to new or unexpected facts is fundamental in the process of environmental planning. To have amplitude of sights and being able to include as many perspectives as possible, shall influence positively the quality of the results obtained by the analyst.

[iii] Method: It assesses the method by which alternatives, criteria and hypotheses have been structured. These could be determined independently by the analyst, given by the political surroundings or defined through a more exhaustive analysis of the *problematique*, such as an institutional Analysis.

[iv] Verification: whether the analyst's job has undergone some type of process of control. It ranges from external (i.e. academic sphere, experts) to occasional or unknown.

[v] Colleague consensus: the degree of acceptance by the rest of the experts working in the same field.

[vi] Extended Peer Acceptance: level of acceptance by the rest of actors involved in the DMP will be considered. Occasionally, in planning processes, stakeholders may not accept the opinion of a specific expert. This lack of credibility could affect negatively the planning process generating a situation of instability.

[vii] *Methods of generation of Information*. Enables the analysis of the 'origins' of the information that will be used during the planning process. It ranges from the experimental sort (experimentation) with a small and controlled degree of uncertainty to the estimations made by experts. It is necessary to stress the importance of the 'extended facts' where the opinions and perceptions of the different stakeholders are displayed. It corresponds to a degree of uncertainty smaller than that derived from the use of historical series of data; this is due to the fact that we are dealing environmental issues that have characteristics of novelty and complexity (it is necessary the internalisation of the perceptions of the involved agents.

[vii] *Information generators*. Specifies the type of information being analysed, ranging from experts' knowledge (scientific knowledge), Public Institutions' position, information generated by stakeholders (consultancy), opinions of

the different agents involved in the issue and the perceptions of the population at large about the relevant issue (perceptions of the population).

[vii] *Verification.* The source of the information is considered here as relevant as the knowledge of whether the data have been checked. It ranges from an external control considered as assuring a greater reliability, to situations in which such control does not exist or it is unknown (none).

[viii] *Colleague Consensus.* It reflects the degree of acceptance of the analysed information by the rest of researchers working in the same field.

[ix] *Extended Peer Acceptance.* Whereas the previous category reflects the degree of acceptance of the information by the expert's own community, here it is evaluated the level of acceptance by the rest of actors involved in the process.

[x] *Source of Legitimacy.* The purpose of this criterion is to specify the legitimising framework of he information. In many cases the credibility of the data is closely related to the Institution that generates or sanctions it. Several distinctions are used, from the academic or political context to perceptions shaped by traditions and cultural heritage of the community involved (generally speaking, the local knowledge).

[xii] *Fitness for purpose:* This category is a bridge between the analysis of the applicability of the information and its related uncertainty. The adjustment degree reflects whether the available information has been developed specifically for the problem at hand or rather it was developed for other cases. An analysis of the degree of connectivity between the information available and the relevant *issue* is necessary. This is done through categories ranging from complete fitness (taking place when the information is developed specifically for the specific problem) to weak fitness (when the information used has been generated in other circumstances where similar problems have been addressed). Scales of low adjustment apply, when the applied data come from studies about other different issues.

[xii] *Relevanc*e Evaluates the relevance of this information in relation to the case study (how the available information corresponds to the relevant issue).

[xiii] *Completeness.* Establishes the level of incompleteness of data series or aspects of the case study in which data or knowledge are not controlled (or controllable), or are even incomplete.

[xiv] *Accessibility.* Shows the degree of accessibility to both the information used and the information needed in the planning process. It is an important aspect in the quality assurance of these types of studies where there is a high dependency on the availability of the information. Thus, if the information is not openly available (or available to a restricted audience) the entire process (review, e.g.) will be affected. This category is divided into five types ranging from the highest availability and accessibility to the lowest where the information is unavailable.

[xv] *Intelligibility.* It enables the analysis of the level of applicability of the information in its use. Intelligibility is defined as the level of training needed by the agents to be able to interpret (and manipulate, perhaps!) this information. It ranges from a total understanding of the information by all the agents implicated in the process (actors involved) to a situation where only the experts in the field (to which that information pertains) can consider/understand it and analyse it (experts). Between those two extremes, three more situations are contemplated: Those consist of the politico-institutional, the consultancy and the academic contexts.

[xv] *Methods of generation of Information.* Enables the analysis of the 'origins' of the information that will be used during the planning process. It ranges from the experimental sort (experimentation) with a small and controlled degree of uncertainty to the estimations made by experts. It is necessary to stress the importance of thc 'extended facts' where the opinions and perceptions of the different stakeholders are displayed. It corresponds to a degree of uncertainty smaller than that derived from the use of historical series of data; this is due to the fact that we are dealing environmental issues that have characteristics of novelty and complexity (it is necessary the internalisation of the perceptions of the involved agents.

[xvi] *Intelligibility.* It enables the analysis of the level of applicability of the information in its use. Intelligibility is defined as the level of training needed by the agents to be able to interpret (and manipulate, perhaps!) this information. It ranges from a total understanding of the information by all the agents implicated in the process (actors involved) to a situation where only the experts in the field (to which that information pertains) can consider/understand it and analyse

it (experts). Between those two extremes, three more situations are contemplated: Those consist of the politico-institutional, the consultancy and the academic contexts.

xvi *Methods of generation of Information.* Enables the analysis of the 'origins' of the information that will be used during the planning process. It ranges from the experimental sort (experimentation) with a small and controlled degree of uncertainty to the estimations made by experts. It is necessary to stress the importance of the 'extended facts' where the opinions and perceptions of the different stakeholders are displayed. It corresponds to a degree of uncertainty smaller than that derived from the use of historical series of data; this is due to the fact that we are dealing environmental issues that have characteristics of novelty and complexity (it is necessary the internalisation of the perceptions of the involved agents.

xvi *Information generators.* Specifies the type of information being analysed, ranging from experts' knowledge (scientific knowledge), Public Institutions' position, information generated by stakeholders (consultancy), opinions of the different agents involved in the issue and the perceptions of the population at large about the relevant issue (perceptions of the population).

xvi *Verification.* The source of the information is considered here as relevant as the knowledge of whether the data have been checked. It ranges from an external control considered as assuring a greater reliability, to situations in which such control does not exist or it is unknown (none).

xvi *Colleague Consensus.* It reflects the degree of acceptance of the analysed information by the rest of researchers working in the same field.

xvi *Extended Peer Acceptance.* Whereas the previous category reflects the degree of acceptance of the information by the expert's own community, here it is evaluated the level of acceptance by the rest of actors involved in the process.

xvi *Source of Legitimacy.* The purpose of this criterion is to specify the legitimising framework of he information. In many cases the credibility of the data is closely related to the Institution that generates or sanctions it. Several distinctions are used, from the academic or political context to perceptions shaped by traditions and cultural heritage of the community involved (generally speaking, the local knowledge).

xvi *Fitness for purpose:* This category is a bridge between the analysis of the applicability of the information and its related uncertainty. The adjustment degree reflects whether the available information has been developed specifically for the problem at hand or rather it was developed for other cases. An analysis of the degree of connectivity between the information available and the relevant *issue* is necessary. This is done through categories ranging from complete fitness (taking place when the information is developed specifically for the specific problem) to weak fitness (when the information used has been generated in other circumstances where similar problems have been addressed). Scales of low adjustment apply, when the applied data come from studies about other different issues.

xvi *Relevancy.* Evaluates the relevancy of this information in relation to the case study (how the available information corresponds to the relevant issue)

xvi *Completeness.* Establishes the level of incompleteness of data series or aspects of the case study in which data or knowledge are not controlled (or controllable), or are even incomplete.

xvi *Accessibility.* Shows the degree of accessibility to both the information used and the information needed in the planning process. It is an important aspect in the quality assurance of these types of studies where there is a high dependency on the availability of the information. Thus, if the information is not openly available (or available to a restricted audience) the entire process (review, e.g.) will be affected. This category is divided into five types ranging from the highest availability and accessibility to the lowest where the information is unavailable.

xvi *Intelligibility.* It enables the analysis of the level of applicability of the information in its use. Intelligibility is defined as the level of training needed by the agents to be able to interpret (and manipulate, perhaps!) this information. It ranges from a total understanding of the information by all the agents implicated in the process (actors involved) to a situation where only the experts in the field (to which that information pertains) can consider/understand it and analyse it (experts). Between those two extremes, three more situations are contemplated: Those consist of the politico-institutional, the consultancy and the academic contexts.

[xvi] *Methods of generation of Information.* Enables the analysis of the 'origins' of the information that will be used during the planning process. It ranges from the experimental sort (experimentation) with a small and controlled degree of uncertainty to the estimations made by experts. It is necessary to stress the importance of the 'extended facts' where the opinions and perceptions of the different stakeholders are displayed. It corresponds to a degree of uncertainty smaller than that derived from the use of historical series of data; this is due to the fact that we are dealing environmental issues that have characteristics of novelty and complexity (it is necessary the internalisation of the perceptions of the involved agents.

[xvi] *Information generators.* Specifies the type of information being analysed, ranging from experts' knowledge (scientific knowledge), Public Institutions' position, information generated by stakeholders (consultancy), opinions of the different agents involved in the issue and the perceptions of the population at large about the relevant issue (perceptions of the population).

[xvi] *Verification.* The source of the information is considered here as relevant as the knowledge of whether the data have been checked. It ranges from an external control considered as assuring a greater reliability, to situations in which such control does not exist or it is unknown (none).

[xvi] *Colleague Consensus.* It reflects the degree of acceptance of the analysed information by the rest of researchers working in the same field.

[xvi] *Extended Peer Acceptance.* Whereas the previous category reflects the degree of acceptance of the information by the expert's own community, here it is evaluated the level of acceptance by the rest of actors involved in the process.

[xvi] *Source of Legitimacy.* The purpose of this criterion is to specify the legitimising framework of he information. In many cases the credibility of the data is closely related to the Institution that generates or sanctions it. Several distinctions are used, from the academic or political context to perceptions shaped by traditions and cultural heritage of the community involved (generally speaking, the local knowledge).

[xvii] *Intelligibility.* It enables the analysis of the level of applicability of the information in its use. Intelligibility is defined as the level of training needed by the agents to be able to interpret (and manipulate, perhaps!) this information. It ranges from a total understanding of the information by all the agents implicated in the process (actors involved) to a situation where only the experts in the field (to which that information pertains) can consider/understand it and analyse it (experts). Between those two extremes, three more situations are contemplated: Those consist of the politico-institutional, the consultancy and the academic contexts.

[xviii] *Methods of generation of Information.* Enables the analysis of the 'origins' of the information that will be used during the planning process. It ranges from the experimental sort (experimentation) with a small and controlled degree of uncertainty to the estimations made by experts. It is necessary to stress the importance of the 'extended facts' where the opinions and perceptions of the different stakeholders are displayed. It corresponds to a degree of uncertainty smaller than that derived from the use of historical series of data; this is due to the fact that we are dealing environmental issues that have characteristics of novelty and complexity (it is necessary the internalisation of the perceptions of the involved agents.

[xix] *Information generators.* Specifies the type of information being analysed, ranging from experts' knowledge (scientific knowledge), Public Institutions' position, information generated by stakeholders (consultancy), opinions of the different agents involved in the issue and the perceptions of the population at large about the relevant issue (perceptions of the population).

[xx] *Verification.* The source of the information is considered here as relevant as the knowledge of whether the data have been checked. It ranges from an external control considered as assuring a greater reliability, to situations in which such control does not exist or it is unknown (none).

[xxi] *Colleague Consensus*. It reflects the degree of acceptance of the analysed information by the rest of researchers working in the same field.

[xxii] *Extended Peer Acceptance*. Whereas the previous category reflects the degree of acceptance of the information by the expert's own community, here it is evaluated the level of acceptance by the rest of actors involved in the process.

[xxiii] *Source of Legitimacy*. The purpose of this criterion is to specify the legitimising framework of he information. In many cases the credibility of the data is closely related to the Institution that generates or sanctions it. Several distinctions are used, from the academic or political context to perceptions shaped by traditions and cultural heritage of the community involved (generally speaking, the local knowledge).

[xxiv] *Fitness for purpose:* This category is a bridge between the analysis of the applicability of the information and its related uncertainty. The adjustment degree reflects whether the available information has been developed specifically for the problem at hand or rather it was developed for other cases. An analysis of the degree of connectivity between the information available and the relevant *issue* is necessary. This is done through categories ranging from complete fitness (taking place when the information is developed specifically for the specific problem) to weak fitness (when the information used has been generated in other circumstances where similar problems have been addressed). Scales of low adjustment apply, when the applied data come from studies about other different issues.

[xxv] *Relevancy*. Evaluates the relevancy of this information in relation to the case study (how the available information corresponds to the relevant issue)

[xxvi] *Completeness*. Establishes the level of incompleteness of data series or aspects of the case study in which data or knowledge are not controlled (or controllable), or are even incomplete.

[xxvii] *Accessibility*. Shows the degree of accessibility to both the information used and the information needed in the planning process. It is an important aspect in the quality assurance of these types of studies where there is a high dependency on the availability of the information. Thus, if the information is not openly available (or available to a restricted audience) the entire process (review, e.g.) will be affected. This category is divided into five types ranging from the highest availability and accessibility to the lowest where the information is unavailable.

[xxviii] *Intelligibility*. It enables the analysis of the level of applicability of the information in its use. Intelligibility is defined as the level of training needed by the agents to be able to interpret (and manipulate, perhaps!) this information. It ranges from a total understanding of the information by all the agents implicated in the process (actors involved) to a situation where only the experts in the field (to which that information pertains) can consider/understand it and analyse it (experts). Between those two extremes, three more situations are contemplated: Those consist of the politico-institutional, the consultancy and the academic contexts.

[xxix] *Methods of generation of Information*. Enables the analysis of the 'origins' of the information that will be used during the planning process. It ranges from the experimental sort (experimentation) with a small and controlled degree of uncertainty to the estimations made by experts. It is necessary to stress the importance of the 'extended facts' where the opinions and perceptions of the different stakeholders are displayed. It corresponds to a degree of uncertainty smaller than that derived from the use of historical series of data; this is due to the fact that we are dealing environmental issues that have characteristics of novelty and complexity (it is necessary the internalisation of the perceptions of the involved agents.

[xxx] *Information generators*. Specifics the type of information being analysed, ranging from experts' knowledge (scientific knowledge), Public Institutions' position, information generated by stakeholders (consultancy), opinions of the different agents involved in the issue and the perceptions of the population at large about the relevant issue (perceptions of the population).

[xxxi] *Verification*. The source of the information is considered here as relevant as the knowledge of whether the data have been checked. It ranges from an external control considered as assuring a greater reliability, to situations in which such control does not exist or it is unknown (none).

ANNEX IV

Problems in Contingent Valuation

1 Dichotomous Choice versus Open-ended Formats

A key decision in designing a CVM survey is how to ask the WTP question. A variety of formats are available including: open-ended, dichotomous choice, double or triple bounded dichotomous choice, payment card or payment ladder. The two most widely used formats, open-ended and dichotomous choice, are discussed here.

Early CVM surveys used the open-ended elicitation method (Hammack and Brown 1974), which is still widely used. An open-ended question format was thought useful as the researcher obtains respondents actual WTP and not an indicator of it, as derived from many of the other formats. It also allows the use of relatively straightforward statistical techniques in analysis (Mitchell and Carson 1989). Despite its popularity this format has been criticised. Some analysts have claimed answering an open-ended question is an "extremely difficult task" (Bennett and Tranter 1998: 536) and because of this, the format may produce more non-responses or protest bids than other methods (Mitchell and Carson 1989). Respondents may therefore want information about the cost of the project, which as far as economic theory is concerned is irrelevant tot their valuation, and which can bias the result by giving leading information (e.g., the aim is to find WTP for a car; how much it cost to produce is irrelevant). A number of authors have suggested that an open-ended format will encourage strategic behaviour, in particular free riding, when voluntary contributions are the payment mechanism used (Arrow et al. 1993; Ready, Buzby and Hu 1996; Bateman, Langford and Rasbash 1999; Carson, Groves and Machina 1999). Others have suggested that responses to open-ended questions may be extremely sensitive to trivial characteristics presented in the scenario (Ready, Buzby and Hu 1996). Finally, although researchers expected that open-ended bid response formats would create very high WTP estimates, this has not proved to be the case (Carson, Groves and Machina 1999). Respondents may choose a bid with which they feel comfortable, which some claim is likely to be less than their maximum WTP (Brown, Peterson and Tonn 1995).

Dichotomous choice formats are the most widely used alternatives to open-ended questions. Although the dichotomous choice format had been used for many years, interest has increased since the National Oceanic and Atmospheric Administration (NOAA) panel in 1993 recommended its use over other elicitation formats (Arrow et al. 1993). The dichotomous choice method was first developed by Bishop and Herberlein (1979); they called it the take it or leave it approach. This approach predefines a number of possible maximum WTP amounts. Each respondent is then asked their WTP just one of these amounts i.e., each predefined amount is administered to a sub-sample of respondents. The selection of bids is usually determined by a previously conducted open-ended questionnaire so that the WTP distribution can be ascertained and the bids for the dichotomous choice questionnaire selected from this range (Bowker and Stoll 1988; Mitchell and Carson 1989; Rollins 1997). The method is described as having a number of economically desirable properties. First, it is "incentive compatible" meaning that the respondent is encouraged to reveal their preferences for/against the change in question i.e., without free riding or strategic behaviour. Second, the respondents' task is regarded as simple in that they make a judgement about a single price only rather than determining a figure unaided or considering a range of different prices. Third, this is said to be the type of decision that a consumer faces in actual market transactions (Bennett and Tranter 1998), although this denies institutional variation in actual markets, shopping around, and the ability to haggle.

A range of problems are recognise with respect to the dichotomous choice elicitation format. The approach provides only an indicator of WTP, not an actual value and requires a large number of responses so that statistical methods can be used to calibrate WTP estimates. Willis (1995) suggests over 2,000 responses are required which means that it is more costly than open-ended methods. The statistical techniques required can be complex and the results of such analysis may be particularly sensitive to functional form, truncation and the statistical estimator of WTP adopted (Bowker and Stoll 1988; Mitchell and Carson 1989; Hanemann and Kanninen 1999). Furthermore, the dichotomous choice format suffers from what has been called "yea-saying", which causes WTP estimates to be biased upwards. Yea-saying is where the respondent readily agrees to the proposed price even when they are not prepared to pay it in reality. This may partly account for the finding that dichotomous choice formats reveal higher WTP estimates than open-ended formats (Boyle et al. 1996). The initial values still have to be derived, usually by carrying out a pilot open-ended CVM study so that a dichotomous choice format implicitly includes all the problems of open-ended formats. Also as the initial values determine the outcome there is an anchoring bias introduced at this stage.

The NOAA panel recommended the use of dichotomous choice (Arrow et al. 1993), causing CVM researchers to move away from open-ended elicitation formats. However, the use of dichotomous choice conflicts with another NOAA recommendation that the most "conservative" (i.e. lowest value) questionnaire design should be utilised, whereas dichotomous choice is know to produce higher values than open-ended formats. This recommendation of "conservative" design has itself been criticised as theoretically unsound as well as politically motivated (Knetsch 1994). Ready *et al* (1996: 409) suggests that those conducting CVM research may have been "too quick to abandon continuous methods in favour of the dichotomous choice method". Due to the increasing knowledge of problems with dichotomous choice, and the lack of clarity as to what is most appropriate, there appears to have been a resurgence in the popularity of the open-ended approach. This is seen as especially suitable to certain situations and non-American contexts. Bennet and Tranter (1998) suggest that open-ended formats may be most suitable when personal interviews are used and when the respondent is familiar with the good. The extent to which the institutional context of the USA can be generalised is limited and the acceptance of markets as well as referenda and their structure is also culturally specific. Hence Willis (1995) has noted that voting on hypothecated taxes is totally outside UK experience.

Payment Vehicle

The payment vehicle is the means by which the respondent is asked to make her bid. A number of different payment vehicles have been used in CVM questionnaires such as taxes (Department of the Environment Transport and the Regions 1999), charitable donations (Macmillan, Smart and Thornburn 1998), entrance fees (Willard 1980; Kenyon, Nevin and Hanley 2001), and trust funds (Spash 2000). Mitchell and Carson (1989) make a number of recommendations about the use of payment vehicles in CVM questionnaires; that the payment vehicle should be familiar to the respondent, that it should have a plausible link to the environmental good being valued, that it should be neutral, in that it doesn't encourage a particular response, unless a policy is being evaluated and that the payment vehicle should be incentive compatible. This means that the vehicle ensures that it is in the respondent's own selfish interest to reveal their preferences for the good in question, i.e. it does not encourage free riding or strategic behaviour.

Information Provision

In a hypothetical market, respondents combine information provided to them regarding the good to be valued, and how the market will work, with information they already hold on that good. Their responses may be influenced by either hypothetical market or commodity specific information given to them in the survey. This phenomenon implies that WTP/WTA values are endogenous to the valuation process. Ajzen et al. (1996) concluded that the information provided in CVM surveys can profoundly affect WTP estimates, and that subtle contextual cues can seriously bias these estimates under conditions where the good is of low personal relevance. However Randall (1986) has argued that CVM answers should vary under different information sets, otherwise the technique would be insensitive to significant changes in commodity framing. The divergence of opinion here relates to information provided which is regarded as important to the decision process and should therefore have an impact but fails to do so, and information which would be regarded as peripheral or irrelevant but which does have an impact on stated behaviour. Indeed, the effects of information may be inappropriately labelled as bias, depending on the way in which WTP/WTA is changed. Information which improves the knowledge of an individual concerning the characteristics of a good can be regarded as informing a consumption decision. Information which alters preferences is more problematic for the economic model which assumes preferences are given. Individual preferences can hardly be regarded as exogenous to the valuation process when the goods are unfamiliar and/or never traded in a market. The idea of a neutral set of information has been shown empirically to be false with some respondents being informed while others have their preferences formed by the same information (Spash 2002). This work also shows the importance of ethical positions with regard to how information is treated. Researcher's judgement in providing information is unavoidable, but can be aided by the use of focus groups and pilot survey testing.

Embedding

Related to issues of information provision in CVM is that of embedding. The embedding problem arises when the component parts of an individual's valuation are evaluated separately and when summed found to exceed the valuation placed upon the whole. This has been attributed by some to valuation of the moral satisfaction from contributing to a worthy cause ('warm glow') rather than the good itself (Kahneman and Knetsch 1992). The counter reaction has been that CVM surveys finding embedding are flawed in some way which creates the part-whole bias, and that this can be corrected by careful survey design (Carson and Mitchell 1993; Hanemann 1994). However Bateman *et al.* (1997) have provided experimental evidence for the existence of part-whole bias for private goods outside of the CVM context. They therefore suggest the problem lies with economic preference theory rather than the CVM approach. Despite this careful design of the questionnaire is required to ensure the respondent is aware of the boundaries of the entity being valued and testing for this within the questionnaire can be useful.

Going Beyond Institutional Guidance

In addressing the type of concerns outlined above, many researchers have looked to authoritative guidance (e.g., Mitchell and Carson 1989) and government agencies have commissioned their own reports (e.g., NOAA 1993; Bateman et al. 2002) which prescribe standard protocols for "best practice". These guides tend to recommend a whole range of ideal approaches personal interviews, random sampling, full information on the resource change and checks for understanding of the information given, careful design and pre-testing of questionnaires, reinforcing budget constraints and so on. Most twenty-minute surveys are unable to address what prove to be a long list of prescriptions some of which lack theoretical justifications (e.g. "conservative" design, blanket use of WTP) and others of which remain highly contentious (e.g. protocols for the exclusion of data). Furthermore, generally, these guidelines have failed to address a range of issues related to the CVM raised by non-economists, such as the role of ethics, attitudes and norms and general environmental awareness in formulation of WTP values. At the same time non-economists have failed to comprehend the state of the art in CVM design. The empirical work reported here addresses issues raised in the philosophy and social psychology literature on environmental valuation while conducting a state of the art CVM survey.

Literature Cited

Ajzen, I., T. C. Brown and L. H. Rosenthal (1996). "Information bias in contingent valuation: Effects of personal relevance, quality of information and motivational orientation." Journal of Environmental Economics and Management 30(1): 43-57.

Arrow, K., R. Solow, P. R. Portney, E. Leamer, R. Radner and H. Schuman (1993). Report of the NOAA Panel on Contingent Valuation. Washington, Resources for the Future: 38.

Bateman, I., A. Munro, B. Rhodes, C. Starmer and R. Sugden (1997). "Does part-whole bias exist?: An experimental investigation." Economic Journal 107(441): 322-332.

Bateman, I. J., R. T. Carson, B. Day, M. Hanemann, N. Hanley, T. Hett, M. Jones-Lee, G. Loomes, S. Mourato, E. Ozdemiroglu, D. W. Pearce, R. Sugden and J. Swanson, Eds. (2002). Economic Valuation with Stated Preference Techniques: A Manual. Cheltenham, Edward Elgar.

Bateman, I. J., I. H. Langford and J. Rasbash (1999). Willingness-to-pay question format effects in contingent valuation studies. Valuing Environmental Preferences: Theory and Practice of the Contingent Valuation Method in the US, EU and Developing Countries. I. J. Bateman and K. G. Willis. Oxford, OUP: 511-539.

Bennett, R. and R. Tranter (1998). "The dilemma concerning choice of contingent valuation willingness to pay elicitation format." Journal of Environmental Planning and Managment 41: 253-257.

Bishop, R. C. and T. A. Heberlein (1979). "Measuring values of extramarket goods: Are indirect measures biased?" American Journal of Agricultural Economics 61(5): 926-930.

Bowker, J. M. and J. Stoll (1988). "Use of dichotomous choice nonmarket methods to value the whooping crane resource." American Journal of Agricultural Economics 70: 372-381.

Boyle, K. J., F. Reed Johnson, D. W. McCollum, W. H. Desvouges, R. W. Dunford and S. P. Hudson (1996). "Valuing public goods: Discrete versus continuous contingent valuation responses." Land Economics 72: 381-396.

Brown, T. C., G. L. Peterson and B. E. Tonn (1995). "The values jury to aid natural resource decisions." Land Economics 71(2): 250-260.

Carson, R. T., T. Groves and M. J. Machina (1999). Incentive and informal properties of preference questions. Presented at European Association of Resource and Environmental Economists, Oslo, Norway.

Carson, R. T. and R. C. Mitchell (1993). "The Issue of scope in contingent valuation studies." American Journal of Agricultural Economics 75(5): 1263-1267.

Department of the Environment Transport and the Regions (1999). The Environmental Costs and Benefits of the Supply of Aggregates: Phase 2. London, Department of the Environment Transport and the Regions: 208.

Hammack, J. and G. Brown (1974). Waterfowl and Wetlands: Towards Bioeconomic Analysis. Baltimore, Johns Hopkins Press.

Hanemann, M. and B. Kanninen (1999). The statistical analysis of discrete-response CV data. Valuing Environmental Preferences: Theory and Practice of the Contingent Valuation Method in the US, EU, and Developing Countries. I. J. Bateman and K. G. Willis. Oxford, Oxford University press: 302-441.

Hanemann, W. M. (1994). "Valuing the environment through contingent valuation." Journal of Economic Perspectives 8(4): 19-43.

Kahneman, D. and J. L. Knetsch (1992). "Valuing public goods: The purchase of moral satisfaction." Journal of Environmental Economics and Management 22(1): 57-70.

Kenyon, W., C. Nevin and N. Hanley (2001). "Citizens' juries: An aid to environmental valuation?" Environment and Planning C 19(4): 557-566.

Knetsch, J. L. (1994). "Environmental valuation: Some problems of wrong questions and misleading answers." Environmental Values 3(4): 351-368.

Macmillan, D., T. S. Smart and A. P. Thornburn (1998). The importance of realism to experiments comparing cash and CV charitable donation: The case of the Isle of Eigg trust. Agricultural Economics Society, Reading.

Mitchell, R. C. and R. T. Carson (1989). Using Surveys to Value Public Goods: The Contingent Valuation Method. Washington, D C, Resources for the Future.

NOAA (1993). "Natural Resource Damage Assessment Under the Oil Pollution Act of 1990." Federal Register 58(10): 4601-4614.

Randall, A. (1986). The possibility of satisfactory benefit estimation with contingent markets. Valuing Public Goods: An Assessment of the Contingent Valuation Method. R. Cummings, D. Brookshire and W. Schulze. Totowa, New Jersey, Rowman and Allanheld: 114-122.

Ready, R., J. C. Buzby and D. Hu (1996). "Differences between continuous and discrete contingent valuation estimates." Land Economics 72: 397-411.

Rollins, K. (1997). "Wilderness canoeing in Ontario: Using cumulative results to update dichotomous choice contingent valuation offer amounts." Canadian Journal of Agricultral Economics 45: 1-16.

Spash, C. L. (2000). "Ecosystems, contingent valuation and ethics: The case of wetlands re-creation." Ecological Economics 34(2): 195-215.

Spash, C. L. (2002). "Informing and forming preferences in environmental valuation: Coral reef biodiversity." Journal of Economic Psychology 23(5): 665-687.

Willard, D. E. (1980). "Ecologists, environmental litigation, and forensic ecology." Bulletin of the Ecological Society of America 61(1): 14-18.

Willis, K. (1995). Contingent valuation in a policy context: The National Oceanic and Atmospheric Administration report and its implications for the use of contingent valuation methods in policy analysis in Britain. Environmental Valuation: New Perspectives. K. G. Willis and J. T. Corkindale. Wallingford, CAB International: 118-143.

ANNEX V

Table 1. Population and Sample Statistics

	Sample (%)	Scotland (%)
Gender		
Male	48	48
Female	52	52
Age		
Under 25	20	30
25-34	21	14
35-44	19	15
45-54	18	14
55+	22	27
Occupation		
Managers and senior officials	5.2	8.0
Professional occupations	6.5	7.1
Associate professional and technical occupations	3.7	9.2
Administrative and secretarial occupations	5.1	8.4
Skilled trades occupations	8.8	8.0
Personal services	9.3	4.7
Sales and customer service occupations	6.9	5.7
Process, plant and machine operatives	3.0	6.3
Elementary occupations	5.4	8.3
Housewife/husband	9.6	4.4
Retired	13.3	8.2
Student	9.9	15.5
Unemployed	9.9	6.2
No response	3.4	N/a
Total N	1012	5,062,011

IWA Publishing's authorised EU representative for General Product Safety Regulations is Diane D'Arras, 15 rue Duret, 75116 Paris, France, e-mail: safety@iwap.co.uk.

Printed and bound by CPI Group (UK) Ltd, Croydon, CR0 4YY
13/05/2026
02109556-0011